PEELING BACK THE LABELS

ORGANICS, NON-GMOS AND OUR SUSTAINABLE PATH TO FEED THE PLANET

JOSHUA MILLER, DPH, PHD, MBA

ISBN 979-8-9854103-0-3 (Paperback)

ISBN 979-8-9854103-1-0 (Hardcopy)

ISBN 979-8-9854103-2-7 (Ebook)

Edited by Keith Gordon, Duo Storytelling

Cover and layout by Liaison Design Group

Published by roundtable AG

www.joshuamillerdph.com

To Lindsay – My best friend, my partner, my wife. For viewing every challenge as an opportunity and every wild idea as an adventure.

To Ruby, Jude, Jo Rae, and Elenore – There could not be a prouder father. You inspire me to make the world a better place.

Contents

Introduction

Right from the start, I want to acknowledge that I have an inherent bias on the topic we are about to discuss together. When I was younger, I didn't grow up on a farm, didn't go to school to learn anything about agronomy (initially), and didn't have a single family member who was even remotely involved in food production, but agriculture has since become an integral part of my identity. I am fascinated by the challenges facing agriculture and the solutions we need both today and in the future. I have come to appreciate that it is the one industry we are all deeply connected to, whether we live in sub-Saharan Sudan or in downtown Los Angeles; however, it is also an industry from which many of us, particularly in the developed world, are becoming increasingly detached. And with this growing detachment, we are finding ourselves reliant upon others to inform us about how our food is produced, what we should be eating, and what practices we should all run and hide from. Because of this, there is a growing divide that is placing us into one of two camps: pro-conventional agriculture or pro-organic (and non-GMO) agriculture.

This is where *Food for Thought* comes in. While I did not originally go to school to learn about agronomy, after receiving a master's degree in genetics in 2005 I decided to get out of the lab and into the field to learn more about this thing called farming and how our food is produced. And so began the journey that would bring us together today. I have worked as an agronomist, a precision agriculture specialist, a sales and marketing manager, a technical specialist for crop protection, and even a lead for the digital strategy of a major agchem manufacturer for the United States. I have gone back to school to learn about how the multiple disciplines involved in food production—like plant pathology, entomology, ecology, soil science, and environmental law, to name a few—all intersect to influence modern agriculture. I received two doctorates and recently completed a Master of Business Administration to understand more about how agronomics are influenced by the business side of the industry. All the while, I have learned about the past, present, and potential future of our food production system. The one thing I am sure of at this point is that we should all be encouraged about the innovative and entrepreneurial spirit of those individuals who are involved in this amazing industry.

Unfortunately, over my fifteen years in the field, I have seen a proliferation of misconceptions and fallacies around how we talk—and think—about food. Misconceptions? Fallacies? That's right, that is the way I see a lot of the dialogue surrounding food production these days, particularly in regard to the use of chemicals, genetics, and other technologies used to drive the industry forward. This is the space in which I have worked and dedicated my time and passion, and I have seen firsthand the problems and misunderstandings that arise when conventional agriculture is castigated as "less than" or even harmful compared to non-conventional or organic methods. While I respect the fact that advocates for organic agriculture are focused on reducing potential harms, many of them don't realize how hurtful these indictments can be. I have a wife and four kids at home, and for anyone to think that I am in an industry that doesn't put the health and well-being of the population and the environment at the forefront of everything we do, which includes my family, is sometimes too much to bear.

It would be easy to lash out and take a confrontational stance in support of my work, industry, and passion. That is actually how this book started, as an outlet for me to put my thoughts down on paper to provide some sort of cathartic release. But something occurred to me as the words continued to come and the pages continued to accumulate—much of the misleading narrative about both organic and conventional agriculture is not being spread by activists or radical environmentalists but by a misinformed, yet otherwise well-intentioned, public.

This is not to put the blame on the public...well, not entirely. I am big on personal responsibility, and I think it is everyone's obligation to be well-informed on a subject if you want to be able to contribute to the discussion. But how can we know if we're uninformed? When farm and ranch families comprise less than two percent of the population in the United States and the average American is four generations removed from the family farm, there is not a lot of direct connection to our food production systems. A declining number of us are intimately aware of how the food on our plate actually got there. Couple that with a narrative that is increasingly one-sided and who can blame the average consumer for taking the advice of so-called "experts" on what food we should put into our bodies and what to think of the direction the American food production system is going.

This brings me to a point that has changed the way I think about farming. Why is farming held to some nostalgic standard? Why do we presume that the way things were done fifty, sixty, even one hundred years ago were better than they are today? For some reason, every other industry is expected to innovate, advance, optimize. But not farming. It seems as though farming should be reminiscent of the American Gothic painting of the father and daughter standing in front of the quaint little farmhouse, pitchfork in hand. Actually, that painting was never about reality—in fact, it was born out of the very desire for a return to the good ol' days. The artist, Grant Wood, initially created the painting after driving by a house in Iowa. He added the father and daughter to depict "the kind of people [he] fancied should live in that house." While the piece was created with good intent, the image was seen by Iowans as satirical at best or stereotypical (and insulting) at worst. Today, the romanticized vision of farming held by the public stands in stark contrast to the realities of those who work in the industry.

Let's imagine for a moment if farming were to return to this idealized version circa the early 1900s. To do so, we would have to remove all the agricultural innovations of the last century. Synthetic fertilizers, pesticides, crop breeding, mechanization, precision farming. Gone. Yep, looks like a farm from 1900. Ah, the glory days. Families working the ground, using horses and mules to pull the plow, threshing grain and milking cows by hand. You can almost see the Norman Rockwell painting in your mind. These farm families worked to eat and *maybe* have a little left over to sell so the family could buy some cloth or other necessities for the home.

Now, bear in mind, American farmers in the early 1900s were not trying to produce enough food to meet global demand like we are today, as this would have been a few decades before globalization really started to take off. Rather, these American farmers were primarily striving to produce enough food for the domestic population, which was around 76 million at the time. In 1900, there were over 5.7 million full-time operations farming about 880 million acres of farmland. Using these numbers, we can estimate that it required 11.5 acres to feed a single person, and each farm was producing enough food for only thirteen people.

So what stops us from going back to those days? We must have enough land to allow this same level of production to feed our current population, right? That's a big negative. In 2018, there were approximately 900 million acres in agricultural production, or 20 million more than 1900. The same production level on this increased ground would only feed an additional 1.7 million people. Since the US population has increased by around 250 million and not 1.7 million, I guess those other 248.3 million people are just out of luck. And not to mention the countless people from other countries that rely on our exports today.

Now, before I make too many enemies right out of the gate, what I just did was present something called a "straw man." This is a tactic where a person misrepresents someone's argument so that it is easier to defeat. There is often some truth to the straw man argument, but the proposition is wildly distorted so that you are not really arguing against the same point anymore. Take my example. While the numbers are accurate, the goal of those wanting to return to "traditional farming," as it was called in a 2018 documentary titled *The Biggest Little Farm*, is not this drastic. But too often in public debates surrounding hot topics, both sides look for a way to "win" rather than offering rational arguments and listening to what the other person is actually trying to say. I am guilty of this myself, but we need to move past this if we want to have useful discussions that will drive progress.

That is what I will attempt to do in the pages that follow. We will discuss two major topics that need to be better understood by a public that is detached from agriculture. Agriculture and food production are essential to all our lives and it is time to have an honest discussion about the way our plates are filled. The prevailing arguments tend to favor a return to days gone by, and even condemn the businesses involved in modern agriculture. While I believe there is a lot of good intent, I have seen too often how filmmakers, activists, and everyday consumers present emotional arguments that, while strong on storytelling, are lacking in science and reason.

I hope to present my side of the story on each topic alongside ample research; my goal is to spark discussion and maybe even educate a little bit. I am not claiming I won't infuse emotion into the equation, particularly given my experience and love of agriculture and all the technologies that have enabled advancements in the field. But I can promise that the story we are about to unpack together will be grounded in science and reason. There is middle ground to be found, and together we can find it.

In Part One of this book, we will discuss the perceived benefits of organic food. What does that little label tell you—and should it drive your purchase decisions? I will present an argument for the responsible use of chemicals in our modern food production systems, and along the way, I will attempt to enlighten readers about the history of agricultural pesticides and why we can take comfort that the food we buy in the grocery store, even without the organic insignia, is safe. We will also discuss some of the concerns with organic farming practices before closing with some ideas on the most sustainable path forward.

In Part Two, we will turn our attention to…wait for it…GMOs! (Cue cries of terror from the audience) You likely have heard something about this topic and may already have some opinions about the pros and cons of this technology. Well, we need to dig into the science (and fast!) as scientists across numerous disciplines, from food to medicine, are gaining a better understanding of genetics and finding new ways to drive innovation. We are on the cusp of a genetic revolution, and I think we all need to be well-versed on the facts. We will explore the science, the criticism, and hopefully land somewhere in the neighborhood of understanding how genetic modification techniques are essential to sustainably feeding a growing population.

Whether you are currently involved in the agricultural industry or just interested in how the decisions you make in the grocery store affect the health and wellbeing of your family and the planet, I hope this will be a fun, educational, and thought-provoking ride. I am not trying to add new data to any of these conversations, but rather "put together a diffuse and variable body of data and to shape it into a clear account that [will] not only educate the public but also reach people in authority." That was the statement made by Shirley A. Biggs in the book *Silent Spring Revisited* about what she believed her friend Rachel Carson's role was when she wrote *Silent Spring*. The irony of this quote will become apparent later, but for now, let's talk about farming!

Part One

Organic Food Production

Starting in the 1970s, Americans were encouraged to read the back of the package—you know, the description of what is *actually* in the food! But now it seems like the front of the package has become central to how we make selections in the grocery store. *Organic. All Natural. Free Range. Fair Trade Certified. Non-GMO.* The badges of honor that grace our food packages these days are meant to pull at our emotive buying strings. Why wouldn't you choose food that is "All Natural"? The not-so-subtle assertion is that anything else would be *un*natural, so why on earth would you choose that? For what, to save a few bucks?!

In recent years, this type of thinking has permeated the public discourse and it is time that an alternative point of view is presented. So, let's take time in this first section to review the claims surrounding organic agriculture and organic food production. We'll unpack the misconceptions associated with conventional agriculture and food production systems and come away with a deeper understanding of the underlying science. That way, we can all be more-informed consumers.

I don't know about you, but I am the lucky one in my family that gets to brave the supermarkets and do the grocery shopping for my household. Six mouths to feed (plus a giant dog) means that these trips are often long and arduous as I make my way up and down the aisles, meticulously checking off the list that my wife and I put together. But it is also expensive! Now, depending on where you live, you may have a justified gripe with my use of the word "expensive." Living in the United States means that we have access to abundant, relatively low-cost food sources year-round. As such, we spend less of our total income on food than any other country on this planet. As of 2016, only 8% of the average American's expenditures were spent on food.[1]

Sure, we can complain that the price of strawberries is too high when we attempt to buy them out-of-season, or that we notice changes in the price of milk, eggs, or ground beef. But when we compare how much of our income gets spent in the grocery store, we are one of only six countries with single-digit expenditures on food. The other five are Singapore, the United Kingdom, Canada, Ireland, and Switzerland. Contrast that with the six highest countries—Pakistan, the Philippines, Kazakhstan, Cameroon, Kenya, and Nigeria—where food accounts for more than 40% of all expenditures. Nigeria comes in on top with a staggering 59%! Maybe we should think twice when we feel like causing a scene in the meat aisle because the price of ground beef is a dollar higher this week.

I bring this up because it is critical to understanding why the organic industry is seeing the kind of year-over-year growth it is in many countries, especially in industrialized countries like the United States, Canada, and much of Europe. Simply put, it is because we have the luxury to spend more on our food. I know this is overly simplistic and numerous studies have been put forth to determine what motivates individuals to purchase organic food, but if you look at where the organic industry is thriving, it is in the countries that have higher levels of disposable income. When we look at the consumption of organic food per capita by dollars spent, the top ten countries are Switzerland, Denmark, Sweden, Luxembourg, Liechtenstein, Austria, the United States, Germany, France, and Canada.[2]

In shear dollars spent, the United States has the lion's share of the market with 46% of global organic sales, followed by Germany, France, China, Italy, the United Kingdom, Canada, and Sweden. In total, these eight countries account for roughly 83% of all organic food retail sales.[3] This means that the other 188 countries of the world (or whatever number of countries you believe there are—I honestly didn't know this was such a highly contested debate until researching for this book) only make up 17% of the total organic food sales! We already noted earlier that the United States, United Kingdom, and Canada each spend less than 10% of their total expenditures on food, but others on the list, like France and Italy, still don't top 15%.

It should be relatively uncontroversial, then, that a correlation exists between higher levels of disposable income and increased consumption of organic food. Maybe you can draw some connection between higher disposable income and education, leading to the claim that the real correlation is between more education and increased consumption of organic food, but this may be a stretch. Actually, if you wanted to describe the most likely organic food consumer according to research, it would be a middle-aged, college-educated, employed, high-income-earning female.[4,5]

This may only be of passing interest to you, but it leads us to other important insights into the debate around organic food. There will be different reasons that individuals choose to purchase organic food—such as social media, which plays a significant role in driving consumer behavior—but ultimately, we owe it to ourselves to understand the origins of organic farming and what organic food really is (and is not). Only then can we decide whether we should truly be concerned with the safety of agricultural pesticides and determine if organic food is actually "healthier" for us and better for the environment. Once you uncover the answers to these questions, you will be able to determine if it makes sense to continue supporting and promoting organic agricultural practices and whether the organic products in the supermarket are worth the premium price tag. That is the goal of our first topic.

1 Roser M and Ritchie H (2013). Food prices. OurWorldInData.org. https://ourworldindata.org/food-prices

2 Research Institute of Organic Agriculture (2015). Top 10 countries for per capita organic food consumption. The Atlas. https://theatlas.com/charts/SksCx20dx

3 Slabakova B (2020). Organic food statistics about the 2020 market. Health Careers. https://healthcareers.co/organic-food-statistics/

4 Curl CL, Beresford SA, Hajat A, Kaufman JD, Moore K, Nettleton JA, & Diez-Roux AV (2013). Associations of organic produce consumption with socioeconomic status and the local food environment: Multi-Ethnic Study of Atherosclerosis (MESA). PLoS One, 8(7). https://doi.org/10.1371/journal.pone.0069778

5 Baudry J, Méjean C, Péneau S, Galan P, Hercberg S, Lairon D, & Kesse-Guyot E (2015). Health and dietary traits of organic food consumers: Results from the NutriNet-Santé study. The British Journal of Nutrition, 114(12), 2064–2073. https://doi.org/10.1017/S0007114515003761

Organic Food Production: An Origin Story

— The Path to Modern Agriculture —

Even though the demand for organic food has boomed over the last two decades, with global revenue increasing from $18 billion in 2000 to $105 billion in 2018, the organic agriculture movement is rooted much earlier in the 20th century.[1] But before we go there, let's first travel back even earlier to see the modernization in agriculture that was occurring around the world that set us on the path to where we are today.

Many modern farming practices can be traced back centuries. Take the way that we plant crops today, with a modern corn planter or grain drill. Jethro Tull (no, not the British rock band, the British agricultural pioneer) is credited with inventing the horse-drawn seed drill in 1701. While he did introduce an improved version of the implement that allowed for consistent seed spacing and depth, the origins of the seed drill go all the way back to China in the 2nd century . . . BC! A mere *1,700 years later* it was brought to Italy where it was patented by the Venetian Senate in 1566.[2] Before having access to a seed drill, farmers would broadcast seed by hand on top of the soil and try to lightly work it into the ground. The seed drill allowed for increases in the number of seeds that would emerge and survive until harvest, in addition to providing straight rows that could be cultivated with plows to reduce weed pressure.

What about the way crops are harvested today? We see big, shiny combines driving through wheat fields in mid-summer or through corn and soybean fields in the fall across the country, harvesting crops at a rapid pace. But for much of our agricultural history, the crop harvest was very labor-intensive.

Take wheat, for example. Two centuries ago, the wheat crop first needed to be cut down with a sickle—also known as *reaping*—at a rate of approximately 0.3 acres per day per laborer.[3] While there is evidence of earlier machine reapers from ancient times, what we would consider modern mechanical reapers were introduced at nearly the same time by Obed Hussey and Cyrus McCormick in the 1830s. (It is interesting to note that in agricultural circles, McCormick is often credited with introducing the mechanical reaper, but the US Patent and Trademark Office ruled that the Hussey Reaper was responsible for the creation and subsequent innovation of the implement.)

After reaping, the wheat needed to be threshed to separate the grain from the seed head, which was also a manual process done by using a flail. The threshing machine was invented just before the turn of the 19th century in 1794.[4] The major revolution came later in the 1880s when the mechanical reaper and thresher were combined, hence the name combine! The first combines still relied on actual horse power, but steam engines were eventually used before the turn of the 20th century to self-propel the machines through the fields. According to TheHenryFord.org website, it was the sight of one of these steam-powered, self-propelled combines that inspired Ford as a young boy in Michigan to pursue his desire to design and manufacture automobiles. Ford is quoted as saying, "It was the first vehicle other than horse-drawn that I had ever seen. It was intended to drive threshing machines . . . and was simply a portable engine and a boiler mounted on wheels."

Henry Ford obviously followed this passion and helped pioneer the concept of mass-production in automobiles with the introduction of old "tin Lizzie" in 1908. Interesting nugget, the Ford Model T is still ranked as the 9th best-selling car of all time with somewhere between 15 and 16.5 million units sold between 1908 and 1927.[5] Less known, however, is Ford's influence on the agricultural industry as he introduced mass-produced tractors and popularized the equipment under the brand name Fordson in the early 1900s. According to a 1922 article in the *Chilton Farm Journal*, a whopping 40% of all tractors in the US at the time were Fordson tractors![6]

Another individual worth mentioning is Eli Whitney. You may know him as the inventor of the cotton gin, but his journey into agriculture may be seen as serendipitous as he only ended up on a plantation in Georgia after some misfires in his teaching pursuits. Here, he worked with Phineas Miller, the plantation manager, to learn about the production of cotton on the farm.

Cotton is a very labor-intensive crop, requiring the cotton fibers to be cleaned, or separated, from the seed that is attached to the fibers within the bolls. Unlike the long-staple cotton that was grown in coastal areas that could easily be cleaned with simple rollers, the short-staple cotton that was grown more commonly in the South was exceedingly difficult to clean. This type of cotton required tedious efforts to remove the seeds one plant at a time, often at a rate of only one pound per day per cotton picker.

It is important to note that this subject requires continued conversation and reflection because of its centrality to slavery in the United States. However, I hope you can appreciate the inclusion of it here, even as what some may see as a missed opportunity to discuss broader cultural issues, as simply a representation of agricultural innovation. In response to the challenges associated with cleaning the short-staple cotton, Whitney created a machine—a cotton en-*gin*-e—that would remove the seeds from the cotton fibers at a rate of fifty pounds per day. Whitney patented the invention in 1794 and started a company with Miller to manufacture and service cotton gins.

Machinery and the mechanization of agriculture were certainly becoming more prevalent through the 1800s, but it has only continued to get bigger, faster, and more efficient through the decades. Without question, the mechanization revolution was the first critical step towards modernizing agriculture to how we know it today. The next critical step would come from our growing knowledge of plants and the advancements being made in our understanding of plant nutrition.

food for thought...

Unfortunately for Whitney and Miller, their idea for the cotton gin was quickly pirated and they were forced out of business in just a few short years. However, the story ends well for Whitney as he was able to secure a contract from the US government to produce firearms, required by a threat of war with France in 1797. At the time, gunsmiths manufactured muskets one at a time and the national armories had only produced 1,000 muskets in three years. Additionally, because each firearm was essentially unique, parts would need to be produced "on-demand" if they broke or needed to be replaced. Whitney planned to supply the United States with 10,000 muskets within two years by creating machining tools that could be operated by unskilled workers and produce precisely uniform parts that could be assembled to produce a complete musket. As Whitney stated, "The tools which I contemplate to make are similar to an engraving on copper plate from which may be taken a great number of impressions perceptibly alike." Although it took him ten years instead of two to complete the order, Whitney delivered on his promise and ushered in the concept of mass production and interchangeable parts. For this reason, some have argued that it is Eli Whitney, and not Henry Ford, who should be adorned with the moniker "the father of mass production."

One of the early pioneers in the field of plant nutrition was Justus von Liebig, who discovered the importance of inorganic minerals to plants and the importance of nitrogen as an essential plant nutrient. One of his most lasting contributions was the Law of the Minimum, which is still taught in every intro-level agronomy course as "Liebig's Barrel."[7] Essentially, he proposed that there are numerous "critical" resources needed for plant growth, but the scarcest resource is the limiting factor. He used a wooden barrel as an illustration to make this point. If every stave in the barrel represents a critical resource, and the water in the barrel represents the plant's growth (or yield) potential, the barrel can only be filled to the point of the lowest stave.

Today, when every backyard gardener has access to rows of bagged plant food at any hardware store or garden center, it is hard to imagine that there was a time when we didn't understand the most basic concepts regarding plant nutrition. But it was Liebig and other pioneers in the field who generated the knowledge we take for granted today regarding the essential nutrients needed for plant growth and development.

Around this time, it was also becoming apparent that we needed alternative sources for nutrients outside of animal manure if we were to continue to increase productivity and meet the growing agricultural demand. Thus, the development of synthetic fertilizers began in the late 1800s through the 1900s. One of the most notable discoveries was the development of the Haber-Bosch process in the early 1900s by German chemists Fritz Haber and Carl Bosch. Haber first developed the process to synthesize ammonia by fixing nitrogen gas from the atmosphere, for which he received the Nobel Prize for Chemistry in 1918. Carl Bosch helped to translate these methods into large-scale production, and he too received the Nobel Prize over a decade later in 1931.[8]

To put it into layman's terms, they essentially discovered a mechanism to pull nitrogen from the air and transform it into ammonia gas. While ammonia is used in cleaning products, as a refrigerant gas, and in other industrial processes, roughly 90% of ammonia gas today is used to produce nitrogen fertilizer products. This discovery changed the world by providing an abundant source of one of the most critical plant nutrients needed for food production.

We could continue with more examples, but I think we've painted a decent picture of the major advancements in agricultural production and efficiency. Because of these accomplishments, people were able to leave the farm for the first time and pursue other endeavors. However, as with most advancements, there were critics of the new ways of doing things. Concerns grew over the inorganic fertilizers that were being used to feed the crops, the ability to increase the size of fields, and even the effect that the new "heavy" equipment was having on the soil. To many critics, agriculture needed to return to more "traditional" practices, even during a time that today we may consider the "traditional" era of farming.

— The Genesis of the Organic Movement —

Albert Howard, who is regarded as the pioneer of the organic movement, spent much of his career in India where he led several agricultural research centers to improve food production within the framework of "traditional" farming practices.[9] While Liebig was promoting the Law of the Minimum using his barrel analogy, with a strong focus on mineral nutrition and the ability to supplement for higher yields, Howard was advocating for the "Law of Return," which stated that organic waste materials should be recycled into the soil for nutrition, with particular focus on soil fertility and humus content. The term "organic farming" was first coined by Walter James, 4th Baron Northbourne, in the book *Look to the Land*, an organic farming manifesto published in 1940. It would be further disseminated by Howard's book, *An Agricultural Testament*, also published in 1940.

Closer to home, Jerome Irving (J.I.) Rodale led the charge for organic farming in the US when he established the *Organic Gardening and Farming* magazine (which would later become *Organic Farming*, the most widely read gardening magazine in the world). He also founded The Rodale Institute in 1947 to serve as a working organic research farm located in Pennsylvania. In the US, Rodale is regarded as the father of the modern organic farming movement.

By the 1940s, the organic movement was gaining momentum, and over the next several decades, pioneers in the field would conduct research, publish articles and educational pieces, and advocate for organic practices. In 1972, the movement would formally come together on a global scale through the creation of the International Federation of Organic Agriculture Movements (IFOAM) led by Roland Chevriot, president of the French farmer organization Nature et Progrès. Chevriot would serve as the inaugural president of IFOAM and sent invitations to industry leaders and advocates from around the world to join in Versailles, France for the first congress. Along with four other founding members, Lady Eve Balfour (Soil Association of Great Britain), Kjell Arman (Swedish Biodynamic Association), Pauline Raphaely (Soil Association of South Africa), and Jerome Goldstein (Rodale Press of the United States), IFOAM gained global recognition and grew into one of the most influential organic organizations. Today, *IFOAM – Organics International* has over 800 affiliates from more than 100 countries across six continents. On the IFOAM website, you can read the vision statement of the organization: "We foster the broad adoption of truly sustainable agriculture, value chains, and consumption in line with the principles of organic agriculture."[10]

food for thought...

Who just had the thought, "What in the world is humus? Isn't that made from chickpeas!?" No, that would be hummus. Humus (**hyoo**-muhs) is a component of the organic matter in the soil. Believe it or not, soils are very complex. They are made up of layers called horizons that are designated by a letter. The very top layer is the organic horizon (O), which we will come back to in a moment. After this comes topsoil (A), subsoil (B), parent material (C), and bedrock (R). In some soils—like those found in forests or in particularly old soils—there can also be an eluviated horizon (E), a zone where the minerals and organic matter have leached out, leaving only sand and silt particles. This will fall between the A and B horizons when present. Each horizon has different characteristics, has different colors , can have multiple sublayers, and can vary in their size, or depth.

But the one we want to focus on is at the very top, the O horizon, or organic layer. The term organic in scientific terms simply means carbon-based. I know we have been talking about it in a different context regarding food production, but all living things are organic because we are carbon-based. And that is what this zone refers to: the carbon rich zone that is comprised of decomposing leaf litter and plant roots, living and dead microorganisms, and even earthworm excretion. The largest component of this zone, however, is referred to as humus. This is the fraction of the organic material that has been broken down to its mineral components; it is also what gives the soil its dark brown or black color.

Humus helps to improve the physical properties of the soil and its ability to hold water and nutrients, so we are definitely a fan of humus in the agricultural world!...

But humus can be mineralized, or broken down, by tillage, which is why there is a big push toward no-till agriculture. This is a topic we will examine later when we discuss some of the benefits of conventional agriculture over organic agriculture regarding soil health. Additionally, the tillage process exposes the carbon rich humus to the air, resulting in the release of carbon dioxide into the atmosphere. This has brought global attention to soils as there has been a growing push to mitigate climate change through carbon sequestration. Because of the vast areasof agricultural lands, small increases in the soil organic carbon could result in significant decreases in carbon dioxide.

Well, I know you were probably only looking for a quick definition of humus, but I think you could all call yourselves amateur soil scientists after this "Food For Thought!" .●

The organic movement was seeing coordination at a global scale and a growing advocacy network to promote the benefits of organic agriculture. Still, there was a major challenge that the organic agriculture industry faced—agriculture was (and is) highly regulated and there were no formally accepted standards in the industry to define, oversee, and certify organic food producers. The responsibilities were left up to the states. California took the lead by introducing the California Organic Food Act (COFA) in 1979,[11] formally providing legal definitions for organic practices in the state. However, COFA did not provide a mechanism for support or enforcement at a national level. Organic laws and programs were passed in twenty-four more states over the next decade but remained highly decentralized. The result was increasing confusion because the term organic meant different things based upon where you were in the country.

In 1990, the organic agriculture industry received the attention it desired from the federal government with the passing of the Organic Foods Production Act (OFPA). This act was part of the 1990 Farm Bill and, perhaps, one of the most pivotal moments for the US organic industry. According to the Sustainable Agriculture Research and Education (SARE) orgnanization, the OFPA "mandated that the USDA develop and write regulations to explain the law to producers, handlers and certifiers . . . [and] called for an advisory National Organic Standards Board [NOSB] to make recommendations regarding the substances that could be used in organic production and handling, and to help USDA write the regulations."[12]

After twelve long years of work, the rules and regulations mandated by the OFPA in 1990 were finally implemented in 2002. The USDA Agricultural Marketing Service created the National Organic Program (NOP) to unify organic farmers across the country under a single organic certification program. Essentially, this program paved the way to allow foods to be marked with the all-important "USDA Certified Organic" label and provided the framework for the NOSB, which is "comprised of fifteen volunteers from across the organic community: four organic farmers/growers, three environmental/ resource conservationists, three consumer/public interest representatives, two organic handlers/processors, one retailer, one scientist (toxicology, ecology or biochemistry), and one USDA accredited certifying agent."[13] The NOSB was given the "sole authority . . . to recommend additions to the National List of Allowed and Prohibited Sub-stances . . . assist in the development of standards for substances to be used in organic production, and to advise the Secretary [of Agriculture] on other aspects of OFPA implementation."[14]

As with any legislation, I would recommend reading through the entire National Organic Program bill, along with the subsequent regulations and manuals (especially before bed if you suffer from insomnia), but we will go through a few of the key portions together in the next section. First, though, we should ask the question: Why was the NOP and certification so important to the industry? This is the crux of the issue and I will bold it so it doesn't get lost in the rest of the story. **Organic agriculture yields significantly less than conventional agriculture and requires a premium price to be profitable.** This is relevant because the price premium was only made possible for the organic food industry when the USDA provided them with the official organic "seal of approval." Essentially, the USDA provided the marketing tool that organic farmers required to create the demand for a significantly higher-priced, albeit similar, product.

To illustrate this point, let's look at a review by Dr. Steve Savage in which he compares organic and conventional agriculture production using data collected by the USDA in 2014. It was estimated that an additional 109 million more acres would need to be farmed to maintain the same level of production if all crops were raised organically. That would be equivalent, according to Savage, to "all the parkland and wildland areas in the lower 48 states or 1.8 times as much as all the urban land in the nation."[15] Not only should this reduction in land-use efficiency call into question the idea that organic farming is better for the environment (which we will get into later), but it also highlights the reason that a certification program—and, just as important, a good marketing campaign—is essential to the viability of the organic industry as a whole.

By choosing to implement organic farming practices—and ultimately foregoing the use of the modern advances that have enabled our food production systems to keep up with the demands of the growing global population—farmers are sacrificing productivity for an ideology. This is not to say the ideology is bad or not worth pursuing, but it is a countervailing ideology that needs to be justified if we are to assume it is superior in some way to our more widely accepted agricultural practices.

— What's in a Name? —

We've already seen that the term "organic" in its original context simply meant "carbon-based," so what exactly does this designation mean when it is applied to a food label, and what assurances can we take from it? Let's hold any discussions regarding the actual quality or benefit differences between organic and conventional foods for subsequent sections. For now, let's simply review the requirements for the various organic designations and look at what farmers can use in their organic operations.

The National Organic Program (NOP) documentation can be found, in full, in the 2002 Annual Edition of the Code of Federal Regulations.[16] Under Title 7, Part 205, the NOP consists of seven subparts and eighty-one sections for a total of forty-five pages. Codified in the regulations are guidelines ranging from how farm records must be maintained, to what livestock living conditions can be, to how certifying agents can become accredited. Let's start with the basics of certification for an organic farming operation and the organic products it produces.

To sell or market organic products, an operation must first become certified. While the costs associated with certification are usually nominal, the process can be rigorous. First, there is a three-year transition phase where operations must follow organic practices before they are eligible for certification. While this is meant to ensure that any residues from synthetic pesticides or fertilizer are significantly diminished before the operation can be certified organic, it requires a significant commitment because no products can be marketed as organic during the transition phase. After this period, independent inspectors will go through the records and documentation of the operation, evaluate farming practices and soil fertility, and assess storage and handling systems. Then, once the operation acquires the organic certification, inspectors are required to test at least 5% of the operation's total products every year to ensure they meet organic standards. One caveat is that operations grossing less than $5,000 per year from organic sales are exempt from the certification requirement. While they are still required to follow the guidelines we are about to cover, they do not have to undergo the full certification process.

Three organic designations are outlined in the NOP: 100 percent organic, organic, and made with organic (specified ingredients or food group(s)). The first two are likely what come to mind when you think about the organic aisle of the grocery store. While products labeled "100 percent organic" must contain, as you would imagine, 100% organically produced ingredients, products labeled "organic" must contain at least 95% organically produced ingredients. The bar is still set pretty high for the remaining 5%, as the regulations state that "remaining product ingredients must be organically produced, unless not commercially available in organic form, or must be nonagricultural substances or nonorganically produced agricultural products produced consistent with the National List." This list refers to the National List of Allowed and Prohibited Substances, which we will discuss shortly. But for now, the key takeaway is that both of these organic designations get to use the coveted USDA Organic seal on their packaging.

The third classification is the least stringent and does not permit the use of the USDA Organic seal. Products that contain at least 70% organically produced ingredients can make statements on the package indicating that the product was "made with organic (specified ingredients or food group(s))." The remaining 30% of ingredients are prohibited from being produced with a small number of specific practices, such as using sewage sludge for fertilizer, but broadly speaking can be produced with nonorganic or conventional practices.

The National List of Allowed and Prohibited Substances provides a comprehensive reference of what inputs—whether fertilizers, pesticides, soil amendments, or the like—organic producers can and cannot use in their operations.[17] The List is part of the NOP and is continually amended to add and remove products. Feel free to look through the entire list at your leisure, but please note that even organic farmers can use inorganic or synthetic substances to control pests (like diseases, insects, and weeds) and to provide essential nutrients to crops. Of particular interest is the reliance on heavy metal-containing products like copper sulfate for plant disease control and ferric phosphate for slug control. As we get ready to move into a conversation around synthetic pesticides, keep this in mind as both of these "organic" options can cause toxicity in humans and pets when ingested at elevated levels—and are actually no safer than some synthetic, non-organic alternatives.

1 Blobaum R (2021). Selected organic history milestones. Roger Blobaum: The Organic Movement Past and Present. https://rogerblobaum.com/selected-organic-history-milestones/

2 Lumen Learning (2021). The industrial revolution: New Agricultural Tools. History of Western Civilization II. https://courses.lumenlearning.com/suny-hccc-worldhistory2/chapter/new-agricultural-tools/

3 History Source LLC (2019). Story of farming: Reaping. History Link 101. https://historylink101.com/lessons/farm-city/reaping.htm

4 Hounshell DA (1984). (1984), From the American System to Mass Production, 1800–1932: The Development of Manufacturing Technology in the United States. Johns Hopkins University Press.

5 Thiel W (2019). The 25 best-selling cars of all time. MotorBiscuit. https://www.motorbiscuit.com/best-selling-cars-of-all-time/

6 Chilton Company (1922, May). Fordson tractor sales open big field for equipment and accessories. Chilton Tractor Journal, 8(5), 12.

7 Kyle RA and Shampo MA (2001). Justus von Liebig – Leading teacher of organic chemistry. Mayo Clinic Proceedings: Stamp Vignette on Medical Science 76(9). https://www.mayoclinicproceedings.org/article/S0025-6196(11)62112-5/fulltext

8 Encyclopedia Britannica Editors (2020). Haber-Bosch process. Encyclopedia Britannica. https://www.britannica.com/technology/Haber-Bosch-process

9 Heckman J (2006). A history of organic farming: Transitions from Sir Albert Howard's "War in the Soil" to USDA National Organic Program. Renewable Agriculture and Food Systems 21(3).

10 IFOAM – Organics International (2020). About IFOAM – Organics International. https://www.ifoam.bio/about-us

11 California Certified Organic Farmers (2018). Our history. CCOF. https://www.ccof.org/ccof/history

12 SARE Outreach (2003). Transitioning to organic production: History of organic farming in the United States. Sustainable Agriculture Network. https://www.sare.org/publications/transitioning-to-organic-production/History-of-Organic-Farming-in-the-United-States/

13 USDA National Organic Program (2015). National Organic Standards Board. Agricultural Marketing Service National Organic Program. https://www.ams.usda.gov/sites/default/files/media/NOSB%20Fact%20Sheet.pdf

14 National Organic Program (2013). National Organic Standards Board new member guide. https://www.ams.usda.gov/sites/default/files/media/NOP-NOSB-NewMemberGuide.pdf

15 Savage S (2018). USDA data confirm organic yields significantly lower than with conventional farming. Genetic Literacy Project. https://geneticliteracyproject.org/2018/02/16/usda-data-confirm-organic-yields-dramatically-lower-conventional-farming/

16 National Organic Program, 7 C.F.R. § 205 (2002).

17 National Organic Program, 7 C.F.R. § 205, Subpart G (2021).

Organic Food Production:
The Skinny on Agricultural Pesticides

— Pesticides are Poisons . . . Right? —

The idea that "pesticides are poisons" is one of the main motivations that people cite for choosing organic food. Over 90% of respondents in a survey from Natural Grocers said that avoiding pesticides was the number one reason they buy organic.[1] Therefore, I think it is worth a brief detour into the field of toxicology to attempt to shed more light on this topic. Since most of my schooling has been in the life sciences, I was familiar with the phrase "the dose makes the poison," a key principle in the study of toxicology. I was not aware, however, that this phrase is credited to the sixteenth-century Swiss physician, alchemist, and philosopher of the German Renaissance, Paracelsus— or Philippus Aureolus Theophrastus Bombastus von Hohenheim, as he would have been known to his mother.

To be more precise, his actual quote was, "Solely the dose determines that a thing is not a poison."[2] Paracelsus was attempting to defend the use of inorganic substances in his medicinal research to treat illnesses. Medicine was relatively primitive during this era and his critics believed that the substances he was experimenting with were too toxic to have any therapeutic qualities. However, Paracelsus had extraordinary insight into something that we now take for granted— even substances that may be toxic at high doses can be beneficial at low doses (think chemotherapy, to take just a single example). Today he is considered the "father of toxicology."

This concept is particularly important as we discuss the use of pesticides in modern agriculture. Yes, you could argue that all pesticides used in agriculture are "poisons" at a significantly high dose. And if we strictly view pesticides as potentially harmful synthetic chemicals created by large multinational corporations, it is easy to see why the fear exists. But when it's stated that way, even something as common as aspirin can become suspect.

Now, I acknowledge there are significant differences between agricultural pesticides and aspirin, even as part of an analogy like the one we are going to go through. Aspirin is therapeutic to humans and we can control the dose to which we expose ourselves. Pesticides are not therapeutic to humans, although we could argue that they provide an indirect benefit by increasing food production and availability. And we cannot directly control how much we are exposed to—although, as we will see in subsequent sections, pesticides are tightly regulated to ensure that exposure is significantly low. However, bear with me as we play out this scenario.

If we were to view aspirin through the same lens as we do pesticides, we would say this is a potentially harmful (deadly, actually!) synthetic chemical created by Bayer, a multi-national corporation. But every time we take aspirin for fever or muscle pain, we choose, knowingly or not, to accept Paracelsus' contention that the dose makes the poison. Aspirin is, in fact, deadly to humans. When researchers evaluate the acute toxicity of a chemical—or to put it another way, the immediate toxicity of short but high exposure of a chemical—they come up with a number called the LD_{50}. This stands for the *l*ethal *d*ose that kills *50* percent of a test population, expressed as the amount of chemical per weight of the test subject. So, the lower the value, the less chemical is needed to cause harm, or the more toxic the chemical.

Now, we don't go around making unwitting human subjects ingest piles of harmful chemicals until they start, one by one, dropping dead. No, this is where the use of laboratory animals comes in. Rats and mice are the primary test subjects, accounting for a staggering 95% of all lab animals. There are several reasons to use rodents: they have short life spans and reproduce quickly, so several generations can be evaluated in a relatively short time. They are small and easy to handle, which makes them ideal lab specimens. They aren't as cute and cuddly as Fido. But they are also remarkably similar to humans regarding biology, behavior, and genetics.[3]

So, what about aspirin and its deadliness? The LD_{50} value is 200 mg/kg.[4] Using this lethal dose established in rats, we can extrapolate the lethal dose for a typical human being. According to an article on Healthline.com, the average American male, aged 20-39, weighs 196.9 lbs.[5] (Just a side note, they also state that the average male in the 1960s only weighed 166.3 pounds, which may be something we should focus on in another book! But I digress.) A quick conversion to metric and we get a weight of 89.3 kg. The next step is to simply multiply 89.3 kg by 200 mg/kg to find that the lethal dose of aspirin for a typical American male is 17,860 mg. Since a tablet of aspirin contains 325 mg of the drug, the lethal dose is equivalent to approximately 55 tablets of the common pain reliever. That would be a *really* bad headache!

But back to the actual LD_{50} value of aspirin—how does it stack up to other common chemicals? For comparison, here are a few more LD_{50} values: arsenic, 15 mg/kg; bleach fumes, 850 mg/kg; Vitamin A, 2,000 mg/kg; and table salt, 3,000 mg/kg. It looks like the acute toxicity of aspirin falls somewhere between arsenic and bleach fumes . . . nice! This is a great example of understanding the benefit of a chemical, knowing the suggested dose, and determining that the benefit of the chemical outweighs the risk of harm from overexposure.

Let's look at one more chemical, a pesticide that has gotten a lot of attention recently in the public square: glyphosate. You may know it better by the trade name of Roundup, but the active ingredient has been the most widely used herbicide, or weed killer, in the United States since 2001. It was recently classified by the International Agency for Research on Cancer (IARC) as a group 2A chemical, leading to the designation of "probably carcinogenic to humans," and the predictable, ensuing public outcry. But how does this "dangerous" chemical stack up when we compare it to the acute toxicity of our friend aspirin? It has to be more, right? Not so much. With an LD_{50} of 5,600 mg/kg, it would take *twenty-eight times* the amount of glyphosate to be as toxic as aspirin.[6]

This was a fun exercise, but I hate to tell you that I just threw another "straw man" at you. Again, the numbers are all accurate, but this would be an incredibly dishonest way for me to talk about the safety of pesticides. The concern is not, or at least should not be, how much pesticide you must drink in one sitting to kill you. The concern is about the long-term effects of these products and the cumulative amount that we are exposed to over time. Not sure if it will ever happen to you, but if you are ever in an argument with someone about the adverse effects of pesticides and they start throwing LD50 numbers at you, stop the conversation in its tracks and say that what we really need to be focusing on is chronic toxicity... not acute! Lucky for us, this is exactly how chemical manufacturers and our regulatory agencies think about the risk associated with agricultural chemicals.

— Regulations & the Agrochemical Industry —

We should take comfort in knowing that farm groups, chemical manufacturers, regulatory agencies, and even our elected government officials have a storied history of working together to make sure our food production systems are safe and efficient. Some of the earliest laws passed in this arena were adopted to either protect the farmer or the consumer. The very first pesticide legislation passed was the Federal Insecticide Act in 1910, which aimed to prevent fraudulent products from being introduced by deceptive manufacturers or distributors.

food for thought...

Beginning in 1921, pilots began experimenting with spreading insecticides by airplane to kill the sphynx moth larvae in catalpa trees in Ohio. The idea spread to Louisiana where cotton farmers were dealing with the devastating impacts of boll weevil. In 1925, the first commercial crop-dusting company was formed in Monroe, LA called Huff-Daland Dusters, Inc. While most pesticides applied by airplanes today are liquids, early insecticides were powders, hence the name "crop dusters."

Because agricultural crop dusting is a seasonal business, Huff-Daland looked for opportunities to generate revenue in the off-season. South America seasons are opposite of North America, so they started traveling south in the winter for work and eventually diversified their operations when they received a contract to fly mail over the Andes mountains between Peru and Ecuador.

You may be wondering why I thought you would find this history lesson interesting. It's because in 1928, the company got out of the crop-dusting business and renamed itself Delta Air Service, with first operations offering service from Dallas, TX to Jackson, MS through Shreveport and Monroe, LA. That's right, the number one airline in the world today, by revenue, got its start spreading insecticides on cotton fields in Monroe, LA. Pretty cool, huh. ●

The next major piece of legislation is one that still lives to this day, albeit much amended over the years. It was the Federal Insecticide, Fungicide, and Rodenticide Act (FIFRA) of 1947. Like the previous act, FIFRA was initially introduced to prevent the sale of fraudulent products. This was a collaborative effort between farmer organizations, chemical manufacturers, USDA, and Congress. So why was there a need in 1947 for, essentially, a more robust version of the 1910 Federal Insecticide Act?

I'll pause for a moment for you to think about this one. Seriously . . . think about it for a second. What was happening in 1947?

Well, two years earlier, a little thing called World War II ended. In the six years and two days from the time Hitler invaded Poland on September 1, 1939, to the formal surrender of Japan on September 2, 1945, much of the industrialized world had chemists focused on creating new compounds that could be used for war. One particular chemical would have far-reaching impacts on soldiers and civilians alike: DDT.

The pesticide DDT was first synthesized by an Austrian student, Othmar Zeidler, in 1874, although it didn't receive any attention at the time.[7] It wasn't until the mid-1930s when Swiss chemist Paul Hermann Müller transitioned his research at the chemical company Geigy from dyes and tanning agents to pesticides that DDT would truly be "discovered." Apparently, Müller's original objective was to find an effective treatment against moths that caused damage to clothing. According to an article in the *Journal of Military and Veterans' Health*, Müller took some of his newly synthesized DDT home, dusted a container with it, and observed its ability to kill flies. Even after he wiped the container down with an acetone solvent, flies that entered the DDT-treated container were still killed.[8] At this point, Müller knew he had something special, and DDT was patented in 1940.

Switzerland was suffering from food shortages during this time due to insect infestations in their crops, and DDT proved effective at controlling the damaging potato beetle in field tests. It was also shown to be effective at controlling lice and fleas. The insecticidal properties of this new compound were effective not only against residential pests (like the moths that it was created for) but also crop pests and numerous other insects that were disease vectors. Additionally, it required only small quantities and was shown to have no toxic effects on humans. With these characteristics, DDT caught the attention of medical entomologists in both Great Britain and the United States.

Disease has historically been one of the largest causes of death during wartime. While the medical field had made great advances through the 20th century, disease was still a major problem in World War II. Vaccines were used as a preventative measure to protect the Allied soldiers fighting across much of the European and Mediterranean theatres from diseases like typhus, a vector-borne disease that is transmitted through infected feces of the body louse. However, in the early 1940s, epidemic levels of typhus were being seen in much of the civilian populations and new solutions were needed. Iran, for example, saw 19,000 cases and Egypt saw 90,000 cases in 1943 alone.[9]

By executive order of President Franklin Delano Roosevelt, the United States of America Typhus Commission was formed in 1942 to help combat the disease. With Lt. Col. Charles Wheeler at the helm, and in cooperation with Dr. Fred Soper of The Rockefeller Foundation, the United States began to focus on controlling the vector of the disease in addition to vaccinations. New methods of delousing were implemented using DDT to prevent the spread of this disease. Interestingly, the choice to use a powder-duster to ensure the insecticide coated the inside of clothing was used because the Muslim women in Iran refused to disrobe to receive treatments in front of men. Because of this requirement, the powder-duster method was created and allowed for rapid mass delousing that kept typhus levels in check through all Allied-occupied lands in the Middle East and Africa. It was also critical to save the city of Naples, Italy, which was experiencing a major outbreak in 1943-1944 after the Germans retreated from the city.

Disease was wreaking havoc on soldiers fighting in other parts of the world as well—namely, the tropical regions of the South Pacific. According to one statistic, for every two men that died in battle in the Pacific theatre, another five perished from disease.[10] One of the primary diseases was malaria. In 1942, there were 47,663 cases of malaria treated in the Pacific theater, with an infection rate of 251 cases for every 1,000 troops. While most soldiers did not die from the disease, the symptoms of fever, chills, and weakness rendered infected troops ineffective for battle. In Milne Bay, Papua, it was estimated that 12,000 man-days per month were lost because of soldiers being bedridden from malaria.[11]

Anti-malarial drugs were the first line of attack to curb this disease, but several problems arose. The most effective drug, quinine, was produced primarily in the Dutch-controlled colony of Java. However, when the Nazis occupied the Netherlands, they cut off the supply of this drug to Allied troops. Other drugs proved to have serious side effects, and when combined with the unrealistic expectation to stay on a strict treatment regimen while in the heat of battle, many soldiers ended up susceptible to malaria. DDT had already been brought to the South Pacific to help treat lice after the successes seen in combatting typhus outbreaks, but its effectiveness at killing mosquitoes was quickly realized and numerous methods were developed to treat bodies of water where mosquitoes bred. It was found that treatments remained effective for up to six months. With the use of DDT, along with new anti-malarial drugs and effective education by the Army, malaria rates dropped by 70% from what they were at the start of the war.[12]

Without question, the discovery and synthesis of DDT by Müller saved countless lives during World War II by helping fight off typhus and malaria, and in the decades following the war, DDT continued to save millions of lives and completely eradicate malaria in numerous countries. The citation below, again from the *Journal of Military and Veterans' Health* article, details DDT's success following the war:

"*In the early 1950s, the World Health Organization launched the Global Malaria Eradication Program. South Africa was one of the first countries to use the insecticide in 1946 and within several years, malarial areas had decreased. India's malaria control program saw similar decreases. Between 1953 and 1957, morbidity was more than halved from 10.8 percent to 5.3 percent of the total population, and malaria deaths were reduced almost to zero. After DDT was introduced in Ceylon (now Sri Lanka), the number of malaria cases fell from 2.8 million in 1946 to just 110 in 1961. Taiwan also adopted DDT for malaria control shortly after World War II; in 1945, there were over one million cases of malaria on the island; by 1969, however, there were only nine cases, and shortly thereafter the disease was permanently eradicated from the country. Similarly spectacular decreases in malaria cases and deaths were seen everywhere DDT was used.* "[3]

At this point, many of you may be questioning my apparent "one-sided" discussion of DDT that focuses only on the positive benefits. Yes, there is more that we need to discuss about this chemical, and I promise we will get there. But the sentiment that DDT provided overwhelming good to humanity when it was introduced is a fact that I think would be hard to argue, and for the "discovery of the high efficiency of DDT as a contact poison against several arthropods," Müller was awarded the Nobel Prize in Physiology and Medicine in 1948.

The discovery of DDT also paved the way for the introduction of many other pesticides. Some historians have referred to the years following WWII and the introduction of DDT as The Golden Age of Pesticides. New classes of insecticides were synthesized and one of the first herbicides, 2,4-D, was created in 1944 and is still used to this day. All told, there were nearly 10,000 individual pesticides registered with the USDA under the new FIFRA guidelines by 1952!

The FIFRA regulations were critical to keeping Americans safe from the myriad chemicals being introduced. Companies had to go through a process to register new pesticides, and there were even guidelines to ensure that powder chemicals that could be confused with household items, like flour, were dyed to prevent confusion and accidental ingestions, as seen in an excerpt from a New York Times article from June 26, 1947:

"A bill requiring color in some poisons to lessen the chance of housewives putting bug instead of baking powder into their biscuits became law today. President Truman

signed the measure which tightens a 1910 insecticide control law, bringing rat and weed poisons under the act. It also requires coloring of any dangerous poisons that might be mistaken for flour, sugar, salt and the like, registration of poisons before they go on the market, and warning labels."[14]

Clearly, we were still learning a lot about pesticides during this time, and I mean a lot! For instance, it took a long time for the regulations to address two key issues of pesticide application: *persistence* and *selectivity*. The reason that DDT was so attractive to combat malaria was its length of control. As previously mentioned, single applications would control mosquitoes for up to six months, giving it a remarkable degree of persistence. A broad spectrum of activity was also a desirable feature so that pesticide applications would control the maximum number of insect or weed pests, giving it a broad scope of selectivity.

Both characteristics are highly desirable in a vacuum, but what about the impacts they have on non-target pests? Fast forward to the necessary requirements for pesticide registration today, where characteristics like persistence and selectivity are evaluated from the opposite point of view. We now select pesticides that persist only as long as absolutely necessary so they don't remain in the environment for extended periods of time or in the edible portion of the plant. Likewise, pesticides are chosen to be selective for only the target pests, decreasing the chance of beneficial flora and fauna from being negatively impacted. Just like with dyeing certain pesticides to prevent accidents in the kitchen, regulations have continued to evolve to protect both people and the environment.

But let's not get ahead of ourselves. There are a few other critical points along the timeline that should be mentioned as we talk about the strict regulation of pesticides in the United States and in much of the world today. Remember, this section is meant to give a certain level of comfort about the amount of regulatory oversight that exists in the agricultural pesticide world. We will only highlight legislation in the US, but many analogous regulatory bodies and legislative measures can be found in other countries around the world (such as the Agriculture and Rural Development Directorate-General in the European

Union, for example).

The 1950s saw an increased focus on pesticide production, registration, and application. While the Federal Food, Drug and Cosmetic Act (FDCA) was passed in 1938, granting the US Food and Drug Administration (FDA) authority to oversee the safety of food, drugs, medical devices, and cosmetics, two major amendments were passed in the early 1950s to involve the FDA in pesticide regulation to a greater degree: the Pesticides Control Amendment (PCA) and the Food Additives Amendment (FAA). The FDA, given the power to establish tolerances and ban pesticides if the agency determined that they were not safe, would now be responsible for evaluating pesticides based on human and environmental safety.

food for thought...

If there are any US government and civics buffs out there, you may be ready to call me out for not knowing what I am talking about. Yes, the responsibility for setting pesticide tolerances lies with the Environmental Protection Agency (EPA), not the FDA. But, at the time the PCA was passed, the EPA did not exist. It wasn't until 1970 that Richard Nixon created the EPA, but we're not to that decade yet. So, until then, the FDA was responsible for the whole kit and kaboodle.

Today, the FDA enforces the tolerances established by the EPA, but numerous other agencies and governmental departments are also involved in food production. Look at some of the subagencies of the USDA below:

· Agricultural Research Service

· Animal and Plant Health Inspection Service

· Farm Service Agency

· Food and Nutrition Service

· Food Safety and Inspection Service

· Foreign Agricultural Service

· National Agricultural Statistics Service

· National Institute of Food and Agriculture

One more nugget: the FDA is a subagency of the US Department of Health and Human Services, not the USDA, and the EPA is an independent agency that is overseen by Congress but is headed by an administrator appointed by the President. ...●

Although regulation on the agrochemical industry certainly increased in the 1940s and '50s, it was merely a warm-up for what was about to happen in the next few decades. Cue Rachel Carson. I briefly mentioned her a while back, but I think any book focusing on the history of pesticides needs to unpack the monumental impact Carson had on pesticides and agriculture. It has been argued that her work, and specifically her seminal 1962 novel, *Silent Spring*, was critical in driving the passage of the Clean Air Act of 1963, the Wilderness Act of 1964, the National Environmental Policy Act of 1969, the Clean Water Act of 1972, and the Endangered Species Act of 1972. Even the establishment of the Environmental Protection Agency in 1970 has been attributed in no small part to the movement that Carson helped spark! So, how did this one woman initiate so much change, and who was she?

— Rachel Carson . . . Oh, Rachel Carson —

Where do we even begin? A quick Google search for her name will bring up numerous websites and articles written in prose that can only be described as full of admiration and wonder. While I have strived, and will continue to strive, to remain as objective as possible through this book, it's hard to contain my emotion when discussing the impact that Rachel Carson had on the industry I love. Even though I think that Carson's outsized influence has hurt or denigrated many good, honest, hardworking agriculturalists, I understand that neither she nor the movement she launched did so with any malice. They were trying to make the world a better place, just like me and my fellow agronomists and chemists and geneticists. While I know it is blasphemous to speak anything but praise for Carson, bear with me as we unpack the background of this influential woman and where I think we need to take a more objective look at her work and the impact it has even to this day.

We won't discuss Carson's childhood, or even her early adult years, as these have been documented in numerous articles, books, movies, and the like. But a brief synopsis is that she had a great love of nature, was an avid and talented writer, pursued advanced degrees in biology and received a master's degree from Johns Hopkins University in the Department of Zoology. Her focus was on marine biology, and she began a career with the Bureau of Fisheries, which would eventually combine with the Biological Survey to become the US Fish and Wildlife Service. She continued to pursue her talent for writing and earned money as a freelancer until she published her first book, *Under the Sea-Wind*, in 1941. It wasn't until she published her second book, *The Sea Around Us*, in 1951 that she gained considerable esteem for her writing and became a credible voice as an advocate for scientific learning. She published one more book in 1955, *The Edge of the Sea*, before releasing onto the world, in 1962, the piece that would place her firmly in the history books, *Silent Spring*.

Again, please don't take my brief overview of her first 55 years as a slight to her accomplishments. She should be applauded for her pursuit of knowledge, especially in the field of oceanography and marine biology. She should be held up as an inspiration for providing for her mother after the passing of her father, and eventually caring for her two nieces after the passing of her sister. And she should be admired for her determination to persevere through her battle with breast cancer and continue to fight for what she was passionate about until her final days. But, and this is a big but, we should hold her work to the same level of scrutiny as any other *scientist*, as this is the label with which she is most commonly adorned—and possibly even more so because of how effective she was at influencing others with her words and her work.

So, what was the general thesis for *Silent Spring*? Carson used numerous "examples" of communities where DDT had been applied to draw attention to the adverse effects the pesticide was having on a vast array of wildlife and even humans. It was a rallying cry to question the consequences of chemical usage and to be better stewards of the planet. And it was a call to start an activist movement to question authority and demand answers from the government. According to rachelcarson.org, her book:

> *"suggested a needed change in how democracies and liberal societies operated so that individuals and groups could question what their governments allowed others to put into the environment. Far from calling for sweeping changes in government policy, Carson believed the federal government was part of the problem. She admonished her readers and audiences to ask "Who Speaks, And Why?" and therein to set the seeds of social revolution. She identified human hubris and financial self-interest as the crux of the problem and asked if we could master ourselves and our appetites to live as though we humans are an equal part of the earth's systems and not the master of them."[5]*

So, that is all she was trying to accomplish. No big deal!

The book was originally printed in three installments in *The New Yorker*, which is where then-president John F. Kennedy first read Carson's work. JFK saw the momentum building behind Carson as an opportunity to help drive forward his own "agenda to combat pollution by connecting old-style conservation to the new-style environmentalism that called for the protection of earth, air, and water (and all creatures dwelling therein)."[16] In fact, close family friend and Supreme Court Justice William Douglas (who, fun fact, still holds the record for longest term served by a Supreme Court Justice at 36 years), wrote that *Silent Spring* "is the most important chronicle of this century for the human race. This book is a call for immediate action and for effective control of all merchants of poison." I don't know about you, but labeling the agrochemical companies that were working to advance our farming systems as "merchants of poison" does not exude a sense of objectivity from the writer!

Kennedy followed suit and stated at a press event that a full investigation had been launched to determine whether pesticides cause harm to humans, citing Carson's book. The very next day, Kennedy announced that he had created a special panel as part of the President's Science Advisory Committee (PSAC) to study the health and environmental concerns of pesticide usage. The committee returned a report, titled simply *Use of Pesticides*,[17] with their findings. I have read the report cover to cover, and while I am not used to reading government documents, it was surprising to me that no citations were given for the facts and figures presented within. It was not surprising, however, that the language used throughout the report was very vague and non-committal. The benefits of pesticides were presented, as were potential risks, but the last page is what caught the attention of the public, which was already divided over this issue. The last two sentences of the report read: "Public literature and the experiences of panel members indicate that, until the publication of *Silent Spring* by Rachel Carson, people were generally unaware of the toxicity of pesticides. The government should present this information to the public in a way that will make it aware of the dangers while recognizing the value of pesticides." This put Kennedy and his administration squarely in opposition to the agrochemical industry, and the farming community at large.

You can probably sense my prejudicial tone when writing about Rachel Carson. Obviously, Carson's work was impactful, so what is my issue with it? My issue is that the piece was presented and heralded as something it was not. Was it thought-provoking? Yes. Was it well-written? Very much so. Was it written with the purpose to drive change? Certainly. Was it scientifically accurate and intellectually honest by way of presenting a fair representation of topics with supporting facts? Absolutely not. This is the problem. *Silent Spring* could very well have been presented as a thought-provoking look at the potential side effects from chemical use, both to humans and the environment. It could have raised awareness in the general population that, as we continue to advance in science, we should constantly reflect on what the positive and negative outcomes of advancements have been, as well as looking forward to ensure we continually strive for perfection.

But this was not the case. Quite the contrary. An example of the way this book is regarded can be summed up by the statement Jill Lepore wrote in The New Yorker, "The number of books that have done as much good in the world can be counted on the arms of a starfish."[18] This is *Silent Spring*'s legacy: a piece of *scientific* writing so profound that it impacted the entire world for good.

It is only fitting then to treat Carson's book as any good scientist would when peer-reviewing a scientific publication. What are the facts? Are they substantiated? Is it objective and rooted in truth? Remember, the entire premise of the book is that agricultural chemicals, particularly DDT, were "entering into living organisms, passing from one to another in a chain of poisoning and death," and required a "social revolution" to do away with the "poisons" and the infrastructure that would create and support them. As we will see in the coming pages, DDT was grossly misrepresented in Carson's work. It did not kill untold species; and, in fact, its prohibition very well could have cost hundreds of thousands of human lives. So how then could this book have driven such change in both policy and public perception? Let's dig deep into *Silent Spring* and find out.

There are numerous scientists who have come down on the other side of the argument from Carson and provided in-depth discussion and analysis of her writings. However, I would argue that many of them have fallen into the same trap that Carson did by allowing emotion to drive their writing. Consequently, they have risked their arguments by letting their subjectivity override their objectivity. Not all, but definitely some. Still, let's investigate some of the arguments that have been raised and try to provide a rational look "across the aisle," as they say.

There are three individuals we will highlight that provide a comprehensive analysis of Carson's work when evaluated together. The first is Charles Rubin, professor of political science at Duquesne University. His focus, according to the university website, is "on emerging technologies, such as nanotechnology and artificial intelligence, and those who believe they will allow the redesign of humanity."[19] He also wrote the book, *Conservation Reconsidered: Nature, Virtue, and American Liberal Democracy*, in which he argues that the ideas and motivations of classical conservationists, like Theodore Roosevelt, John Muir, Gifford Pinchot, and Aldo Leopold, have been misunderstood and misconstrued by those viewing conservation through a more contemporary lens.

Rubin's article titled "Reading Rachel Carson," published in *The New Atlantis* in 2012, evaluates the way Carson's *Silent Spring* has been interpreted by both supporters and critics, taking an objective and often critical stance of those who fall on both sides of the argument.[20] The section titled "Selecting the Evidence" is particularly enlightening as Rubin pulls examples from the book to point out numerous instances where the facts are either distorted, stretched, or flat-out fraudulent.

One such example is how Carson links radiation (or "chemicals that imitate radiation"), pesticides, and the increase in early childhood leukemia. She strings these together with support from an "authority," in this case Sir Macfarlane Burnet, who had reported that the increased incidence of leukemia in children was likely from a mutagenic stimulus near birth.[21] Because of the way Carson frames her argument for pesticides causing the increased incidence in leukemia, and then immediately referring to Burnet's study as support, the only way to interpret her claim is that Burnet believes that pesticides are the likely culprit for the increase.

However, upon investigation of Burnet's study, he never even mentions pesticides as one of the potential causal agents. In fact, Burnet reveals that the problem is not observed "in colored persons" so it would be highly unlikely for the cause to be from "some all-pervading element associated with advancing civilization." He speculates on possible causes, like coffee, tea, cigarettes, instant baby formula, and pharmaceuticals for pregnant women, but again, never mentions pesticides because, as Carson argues throughout her book, these would be "pervading elements" and would likely affect people of all races. The omission of Burnet's conclusions from Carson's argument is telling. As Rubin states, "She suppressed it. She bracketed his point with observations that made it clear that his thinking in the article cited moved in the same direction as her own."

Another example Rubin provides is Carson's discussion on the impact of DDT on reproduction. She very clearly states that "the possible effect on human beings is seen in medical reports of oligospermia, or reduced production of spermatozoa, among aviation crop dusters applying DDT." There are a lot of big words in this sentence, but two of the most important are easy to understand: "medical" and "reports." The logical conclusion the reader would take away is that this phenomenon has been observed by numerous studies and the results have shown that DDT causes oligospermia (this just means low sperm count), thereby impacting reproduction. Point proved! But wait, what is the citation for these multiple reports? As Rubin highlights, the citation points to only one "report," which was actually a letter to the editor in the *Journal of the American Medical Association* (*JAMA*). A doctor in Phoenix had written a question to the journal editors to inquire whether the oligospermia that he had observed in three crop dusters in his area could be related to DDT, or specifically xylene, a solvent in the pesticide.

As Rubin writes, the response from the editors at *JAMA* begins, "Neither xylene nor DDT is known specifically to impair spermatogenesis," and continues to suggest that real exposures to xylene would elicit symptoms that would "provoke 'medical comment.'" The suggestion the journal provides to the doctor is to investigate whether there has been repeated low-dose carbon monoxide exposure or whether there are pituitary gland malfunctions. Just so we are all on the same page, Carson references "medical reports" that indicate oligospermia is a "possible effect" of DDT observed in crop dusters applying the chemical, when in fact it was a letter to the editor in which *JAMA* responded by letting the doctor know that DDT and xylene are unlikely to cause the observations in question, and even provides two other likely causes.

These examples highlight one of the problems I see with using this book as a catalyst for change in the agrochemical world. Pesticides, and DDT in particular, are portrayed in a negative light by way of assumed conclusions from supposed scientific studies. This is not to say that valid points are not made throughout her book—some that I promise we will get to!—but it is important to understand that in any honest assessment of Carson's work as a piece of *scientific* writing, the fraudulent portrayal of *scientific* evidence should disqualify it as an authoritative work.

The second commentator worth reviewing is Robert Zubrin. It may seem odd that he weighs in on agricultural pesticides given his pedigree. He is an aerospace engineer by trade, received his PhD in nuclear engineering, and spent much of his career advocating for the exploration of Mars and scenario planning for space exploration. He and colleague David Baker developed the Mars Direct plan, a proposal detailing how a human mission to Mars could be done cost-effectively by using the Martian atmosphere to produce the oxygen, water, and fuel needed for the time on the planet's surface and subsequent return flight.[22]

So how does he enter into this conversation? He wrote the book *Merchants of Despair: Radical Environmentalists, Criminal Pseudo-Scientists, and the Fatal Cult of Antihumanism*, which highlights such figures and topics as Thomas Malthus, eugenics, and anti-nuclear campaigns, ultimately arguing that they restrict human activity and freedom. Rachel Carson falls into this category for Zubrin. While the title of his book may come off as a little provocative to those who don't initially share his beliefs, I find his writing well-researched and thoughtful.

Particularly, it provides a relevant perspective to the debate at hand, specifically his analysis of several of the claims that Carson makes in *Silent Spring*.

One of the most hotly debated claims from the book is that DDT endangered US birds with extinction. After all, this was the impetus behind the title. Carson argued that since the introduction of DDT, bird populations were rapidly declining and would eventually render the *spring*time *silent* as all the birds would be eradicated. Zubrin discusses several points to refute the claim that DDT had a negative impact on birds, including numerous studies based on historical and geographically dispersed bird counts that actually showed an *increasing* number of birds during the DDT years, like the evaluations of osprey in Hawk Mountain, PA and the herring gull in Tern Island, MA. There was even a study initiated due to the concerns raised by Carson's book, the North American Breeding Bird Survey, that began in 1966 and carried on through the end of the 1970s. Here again, investigators saw no obvious increases in bird populations after the banning of DDT.

While the data presented by Zubrin is compelling, one of the most cited references to refute Carson's claims is the annual Christmas Bird Count conducted by the Audubon Society.[23] Zubrin provides a summary where he compares the numbers from 1941 as a baseline before DDT was introduced, with 1960 during the peak DDT usage and before the publication of *Silent Spring*. While this could be considered a valid way to look at the data, I do have some concerns about making this the "proof" that Carson's argument was wrong.

I have dug through the Christmas Bird Count data and found that it is easy to find outlier years, so choosing any one year to look at could be misleading. I pulled data for the American robin from the Audubon Society website to show populations between 1940 and 1976. Why the robin, you ask? Because Carson specifically identifies this common bird as being at risk for extinction. As you can see in the chart provided on the next page, there is no decrease in the population of the bird at any point during this time span. In fact, using all data points between 1940 and 1972 (the year DDT was banned), there was a

20% *increase* in counted bird populations. To be even more accurate, we can remove the outliers by eliminating any points that fall outside of one standard deviation and rerun the analysis to show that the American robin actually saw a population increase of 39% between 1940 and 1972 according to these counts! A far cry from being on the verge of extinction.

While the Christmas Bird Count survey data sheds new light on the potential "stretching" of claims around declines in bird populations, this still does not address the most common complaint levied against the pesticide. It was argued at the time, and I would say agreed upon nearly unanimously today, that bird decline was actually due to the thinning effect the pesticide had on the birds' eggshells. DDT, in fact, now serves as the classic example of **bioaccumulation**— the gradual build-up of a chemical in a living organism over time, and **biomagnification**— the increased concentration of a chemical the higher the animal is on the food chain. According to Encyclopedia.com on the topic of bioaccumulation, they state, matter-of-factly, that "levels of DDT were high enough that the birds' eggshells became abnormally thin. As a result, the adult birds broke the shells of their unhatched offspring and the baby birds died. The population of these birds plummeted." But is this a "matter of fact"?

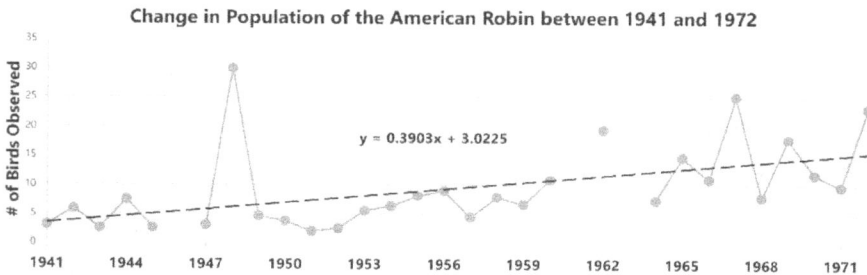

Change in Population of the American Robin between 1941 and 1972

$$y = 0.3903x + 3.0225$$

Chart adapted from the Audubon Christmas Bird Count data. Data outliers that fell more than one standard deviation from the average were removed, and a linear trendline was matched to the data.

There were a significant number of studies from the 1960s through the 1980s to investigate this claim, and I have to be honest, it is hard to reach a definitive conclusion. Those that found evidence of eggshell thinning have had substantial rebuttals to the methods used, admissions that other extraneous factors were also present that could cause eggshell thinning, and questions about the large amounts of DDT needed to cause any effect at all. At the same time, there are ample scientific publications that were unable to find any correlation between DDT and eggshell thinning at all.

Let me provide a specific example. The 1982 *Science* article, "Ban of DDT and subsequent recovery of reproduction in bald eagles," evaluated the viability of bald eagle eggs over the course of sixteen years.[24] Over the course of the entire study, the researcher was only able to collect a total of nineteen addled, or bad, eggs. According to the author, "Statistical comparisons . . . are difficult because of small sample size . . . considerable year-to-year variation, and the fact that eggs were not sampled at random . . . This led to unbalanced sample sizes, and in eight years no addled eggs were found. In addition, the sample consists of eggs that failed and hence is biased." However, even after the previous disclaimer was provided, the author still states that "in spite of these problems . . . I believe that the eggs provide a valid glimpse of contamination levels in the population." So, after a sixteen-year study that the author admits is critically flawed and cannot provide any conclusive results, we are to take his "belief" as the proof that DDT was responsible for reducing the bald eagle population.

Add to this the fact that in 1940, the US Congress passed the Bald Eagle Protection Act because the bird was "threatened with extinction"—the same year DDT was patented and several years before it was available for commercial use. I challenge you to find anything on the internet about bald eagle populations that doesn't briefly mention the 1940 Act before immediately claiming that the use of DDT after WWII poisoned these birds, thinned their eggs, and decimated their populations. Never mind the fact that the publicly available Christmas Bird Count data shows year-over-year increases in bald eagle numbers every year after the passage of the Bald Eagle Protection Act, even during the 1950s and 1960s when DDT was widely used. Regarding the effect DDT had on eggshell thinning, I would argue that the data is inconclusive at best or misrepresented at worst.

Zubrin also spends time refuting claims that DDT was going to threaten life in the oceans. This one cannot actually be attributed to Carson as it was presented after her death in a 1968 *Science* article titled, "DDT reduces photosynthesis by marine phytoplankton,"[25] but it is certainly in line with the scientific outrage against DDT that was generated by Carson's work. The author of the *Science* article was Charles Wurster, whom I find notable because of his position as cofounder of the Environmental Defense Fund, a group that represented the public interest on a total ban of all uses of DDT during an EPA investigation that we will review shortly.

Wurster reported in the publication that DDT in seawater stopped photosynthesis in phytoplankton. That's big, like "DDT = no photosynthesis = no phytoplankton (the foundational producer for all higher order marine life) = no more ocean life" big! This is serious stuff and was taken seriously as well. However, if you look at the experiment, the levels that halted photosynthesis in the phytoplankton was 500 ppb of DDT. I was unaware until reading summaries like this one by Zubrin, but the maximum solubility of DDT in seawater is only 1.2 ppb, a far cry from 500. So, what is going on? To get these extremely high levels, 417 times higher than what is physically possible in nature, Wurster used a solution of saltwater and alcohol to be able to dissolve these elevated levels of DDT. In other experiments that used actual seawater, scientists found no effects on photosynthesis even at maximum saturation levels of DDT.

But never let facts get in the way of a good story! Paul Ehrlich, professor and now president of the Center for Conservation Biology at Stanford University, ran with this narrative. He wrote a dystopian pamphlet in 1969 called *Ecocatastrophe* that described the impending doom from the pesticide revolution and the destruction of the oceans from agrochemical poisons.[26]

An excerpt from the pamphlet reads:

"The end of the ocean came late in the summer of 1979, and it came even more rapidly than the biologists had expected. There had been signs for more than a decade, commencing with the discovery in 1968 that DDT slows down photosynthesis in marine plant life. It was announced in a short paper in the technical journal, Science, but to ecologists it smacked of doomsday. They knew that all life in the sea depends on photosynthesis, the chemical process by which green plants bind the sun's energy and make it available to living things. And they knew that DDT and similar chlorinated hydrocarbons had polluted the entire surface of the earth, including the sea."

Just to put a bow on this point, the argument that high levels of DDT in the ocean could kill algae by stopping photosynthesis is only possible when the seawater is altered beyond its natural state by adding alcohol. Hmm…it sounds like if we avoid the urge to dump trillions upon trillions of gallons of alcohol in our oceans, we should not have to be concerned with DDT killing the algae, a key argument for the case to ban DDT!

Thomas Hughes Jukes provides another point of view in his paper, "DDT, Human Health and the Environment."[27] He breaks down the argument by stating that, even to reach 1 ppb of DDT in the massive volume of ocean water (again, not the 500 ppb that was shown to have a negative effect on algae), "it would take 9,000 years if the total annual production of DDT, 300 million pounds, were dispersed in the oceans each year and there was no breakdown." Sounds like reason enough to ban the chemical to me! Obviously, this is snark on my part, and not an endorsement to be unsafe or reckless with any chemical. But it is hopefully a strong enough point to make you think about digging into the facts of an argument yourself and not simply taking the headlines at face value.

food for thought...

Since Paul Ehrlich was just brought up, I think it is worth taking the time to pause and reflect on how and why we view someone as an authoritative figure or expert if they fall on our side of the argument, while at the same time dismissing others as alarmists, deniers, or even extremists if they fall on the other side of the fence. It seems like we see this at increasing rates today with a growing chasm forming between sides of hot topic debates.

As for Ehrlich, he is the president of a research center at one of the most prestigious universities in the world, so clearly he is viewed as an authoritative figure. But from my perspective, I would say that he is most famous for being horribly wrong in his scientific outlook on population growth and resource limitations. He wrote *The Population Bomb* in 1968, a widely acclaimed book where he predicted that overpopulation would lead to mass famine and societal upheaval in the 1970s and 1980s.[28] "The battle to feed all of humanity is over" and "in the 1970s and 1980s, hundreds of millions of people will starve to death in spite of any crash programs embarked upon now." These are quotes from Ehrlich's book, a required read in many college classrooms.

Meanwhile, there were individuals like Norman Borlaug, heralded as "the father of the Green Revolution," who received the Nobel Peace Prize in 1970 for his work to bring higher yielding, disease-resistant wheat varieties and improved crop management strategies to Mexico, Latin America, and Asia. Based primarily on my experiences, I would put Borlaug in the category of authoritative figure, as he has been credited with saving "more lives than any other person who has ever lived" due to preventing starvation.[29] To contrast the work of these two individuals, at a time when Ehrlich is quoted as saying "India couldn't possibly feed two hundred million more people by 1980," Borlaug was helping find solutions for food shortages . . . and saw India feeding itself within five years of the release of *The Population Bomb*.[30]

The main issue I see is that, even after facts are uncovered and we learn more about the debate topics and arguments at hand, ideology has a way of keeping us anchored to the original ideals of our stance, truth be damned. The idea that population growth was out of control, and therefore was going to put undue strains on the natural resources of the planet, became the essential truth. In fact, Borlaug became a target of environmentalists for now being the cause of increasing populations from the Green Revolution! Not only that, but the methods he pioneered that prevented the purportedly inevitable famine became the enemy because they… required the use of fertilizers and pesticides.

Revisionist history is an amazing thing, but one of Borlaug's former colleagues, Richard Zeyen, professor emeritus in plant pathology at the University of Minnesota, provided a refreshing take from Borlaug himself . "Norman didn't take it personally—usually he just said you had to be there to see the death and starvation and to smell it and taste it. Then he'd ask them, 'What would you do if you knew how to avoid this suffering and death?'"[31]

So, we are left with a problem that I honestly don't know how to solve. I am likely as guilty as anyone for "sticking to my guns" on issues that I care deeply about—this book is probably evidence enough of that fact. I guess all we can do is continually devote ourselves to uncovering the truth, wherever it may lie, and, as hard as it is, conscientiously commit to avoid assigning nefarious intent to others with whom we may not see eye to eye.

J. Gordon Edwards was a professor of entomology at San Jose State University and is the namesake for the entomology museum at the university. He was a staunch proponent for the use of DDT, an equally staunch critic of Rachel Carson, and is the third commentator we will review in this section. You may have heard myths of a guy that ate DDT in public settings to prove how safe it was in the face of the growing backlash after Carson's book. Well . . . it wasn't a myth, it was this guy.[32] Earlier in life, Edwards was an avid mountaineer and naturalist, working as a ranger at Glacier National Park where he was a literal "trailblazer," creating trails to more than 70 peaks within the park and eventually writing the bestselling rock-climbing book, *A Climber's Guide to*

Glacier National Park. He died at the age of 84 while hiking Divide Mountain in the park he loved.[33]

His most concise argument against the ban on DDT was a piece published posthumously in the *Journal of American Physicians and Surgeons* titled, "DDT: A Case Study in Scientific Fraud."[34] As if the title weren't attention-grabbing enough, the abstract starts with a hard-hitting opening:

"The chemical compound that has saved more human lives than any other in history, DDT, was banned by order of one man, the head of the U.S. Environmental Protection Agency (EPA). Public pressure was generated by one popular book and sustained by faulty or fraudulent research. Widely believed claims of carcinogenicity, toxicity to birds, anti-androgenic properties, and prolonged environmental persistence are false or grossly exaggerated. The worldwide effect of the U.S. ban has been millions of preventable deaths."

The "one popular book," if you didn't already know, is a reference to *Silent Spring*. While this article was certainly well-argued, I think Edwards' most compelling piece, and the reason I include him in this book, was his 1992 article published in *21ˢᵗ Century Science & Technology Magazine* titled "The Lies of Rachel Carson."[35] I say the piece is his most compelling because it gives the reader a different perspective about the way Edwards came to "join the detractors of *Silent Spring*." He describes how he was initially "delighted" when the book was first published and how he "eagerly" read the initial condensed version in *The New Yorker*. He was a naturalist working in Glacier National Park and belonged to, as he describes it, "several environmental-type organizations." Even after he started noticing several false statements, Edwards continued to read it with openness. After all, he states, "one can overlook such things when they are produced by one's cohorts, and I did just that."

Hopefully, this statement triggers a moment of inward reflection in all of us. How susceptible are we to the echo chambers we find ourselves in, only hearing the ideas we already believe parroted back to us by other like minds? How able are we to objectively look at facts and keep an open mind as opposed to falling subject to our own confirmation biases and finding the arguments and "facts" that best fit our own ideologies? This was the exact point of the last Food for Thought.

Edwards may be seen as unique in today's culture as he was able to look at the facts being presented, conduct his own research on the topic, and then realize that the arguments being presented were not factual, even if they did support his ideology. He went back through the text, page by page, and took notes on the inaccuracies that were presented. He noticed how Carson was able to word sentences in such a way to imply certain points without actually saying them. He found, repeatedly, where research that was cited in the book to support a point did not actually provide support upon further investigation. He even noted how references were included in the bibliography, not once, but every time they were cited in the book to make the reference section look more robust.

Edwards wrote "The Lies of Rachel Carson" because he became aware that a movie was being made for television to honor Carson and *Silent Spring*. As he states in the article, because he believed "such a movie would further misinform the public, the media, and our legislators, I decided to type up my original rough notes from 1962-1963 and make them available."

We have already reviewed numerous cases of factual misrepresentations in Carson's book, so we don't need to belabor the point any further, but I would encourage any reader to look through Edwards' notes yourself to get a better understanding of all the claims that were taken as fact from *Silent Spring*. After all, we are still dealing with the repercussions of this book today.

Before we move on to the next section, I would like to go back to Zubrin and examine a portion of his essay, adapted from *Merchants of Despair*, where he reviews the actual banning of DDT. Zubrin describes the impact DDT had on the war effort during World War II and the subsequent relief it provided around the world to those suffering from insect-borne diseases, both of which we have already covered at length.

But he then discusses the rise of Carson's book and how it led to the banning of DDT around the globe, giving examples of the mass loss of life that occurred from this decision. One example is in Ceylon, current-day Sri Lanka, where DDT had helped reduce the millions of cases of malaria in the 1940s to only *seventeen* cases in 1963. I already cited this example in the previous section, but what I didn't mention was that after the use of DDT was discontinued in 1964, there was a resurgence of malaria in Ceylon and half a million cases were seen in 1969.

Zubrin quotes the National Academy of Sciences (NAS) that offered support for the use of DDT in a 1970 report that included statements like this:

> "In little more than two decades, DDT has prevented 500 million deaths due to malaria that would otherwise have been inevitable. Abandonment of this valuable insecticide should be undertaken only at such time and in such places as it is evident that the prospective gain to humanity exceeds the consequent losses. At this writing, all available substitutes for DDT are both more expensive per crop-year and decidedly more hazardous."

Even with the endorsement of respected scientific organizations like NAS, a group that publishes the Proceedings of the National Academy of Sciences (PNAS), the second most-cited scientific journal in the world, the momentum from *Silent Spring* proved to be too great. Outside pressure continued to build, and in 1971, the EPA launched a seven-month investigation to determine the fate of DDT.[36]

After reading the entire summary provided at the conclusion of this investigation by the Hearing Examiner, Judge Edmund Sweeney, I was surprised by the findings. Sweeney provides nineteen "Conclusions of Law" at the end of the report before writing his opinion. Several are written in legalese that are a little hard to understand, but some definitely stood out. A few worthy of note:

> 6. The quantum of proof herein is the preponderance of the evidence.

> 9. DDT is not a carcinogenic hazard to man.

10. DDT is not a mutagenic or teratogenic hazard to man.

11. The use of DDT under the registrations involved here do not have a deleterious effect on freshwater fish, estuarine organisms, wild birds, or other wildlife.

12. The adverse effect on beneficial animals from the use of DDT under the registrations involved here is not unreasonable on balance with its benefit.

17. There is a present need for the continued use of DDT for the essential uses defined in this case.

Finally, Sweeney concludes in his written opinion that:

"In my opinion, the evidence in this proceeding supports the conclusion that there is a present need for the essential uses of DDT; that efforts are being made to provide a satisfactory replacement for DDT; and that a co-operative program of surveillance and review can result in a continued lessening of the risks involved."

Well, there it is. The investigation the environmentalists demanded. An impartial hearing that included testimony from 125 expert witnesses and 365 exhibits. So, I guess we would assume that DDT was not banned by the EPA after that conclusive evaluation? We would be wrong. The EPA administrator, William D. Ruckelshaus, took the report, promptly ignored it, and banned DDT in the US in 1972.

Whichever side of the argument you find yourself on, I can guarantee, with almost certainty, that you felt at least a small bit of anger after reading that last paragraph. You were either outraged by the path that Ruckelshaus took, ignoring the result of an independent investigation, or you were incensed by the way in which I presented the facts, a clearly biased portrayal of the events that actually transpired. I have found that opponents of the decision often go straight to corruption or some nefarious motive as the reason for the "injustice." However, supporters of the decision will argue numerous points to support their stance— why banning DDT was a good thing for the planet, why it wasn't really a ban, why it shouldn't matter because DDT was already being phased out, or why resistance to DDT in insect populations was cause enough to ban the pesticide. To be honest, it is hard to separate fact from fiction, especially now since five decades separate us and that fateful day. And that is the reason for this book.

I am clearly on "one side" of the debate. I tend to gravitate toward publications and websites that support my preconceived notions regarding the events at hand. Because of this, my biases are continually reinforced. Even the scientist in me that strives for data and objective truth struggles with the temptation to simply find the echo chamber where "my truth" can be mirrored and amplified. However, when I honestly assess the arguments of those that have a different perspective than myself, I am reminded of a quote by George R. R. Martin in which he states, "Nobody is a villain in their own story. We're all the heroes of our own stories." Just because we may disagree, even vehemently, it is counter-productive to assign motive to others.

I can easily find multiple sources that portray Ruckelshaus as a villain—in the pocket of "Big Environment" and doing the bidding of groups like the Environmental Defense Fund and the Audubon Society. Likewise, I can find articles that portray the decision of Ruckelshaus as nothing short of the obvious, natural conclusion based not only on the EPA investigation, but also the broader understanding of DDT and the available data. It is precisely when these diametrically opposed arguments are strongest that all parties need to take a step back for perspective.

While I have heard the arguments in favor of Ruckelshaus, I continue to disagree with the EPA's decision to ban DDT. Let me share my reasoning with you in an attempt to gain some perspective:

1) **The available data at the time was inconclusive regarding the negative impact of DDT on humans and the environment, but the benefit it provided to society was certain.** An independent investigation was launched to come to a consensus, but the conclusion of the investigation was not followed. In light of the incontrovertible understanding that DDT was saving lives that would otherwise be lost to malaria and other

diseases, the banning of the pesticide was perceived as ideological and damaged the public trust in the institution of the EPA.

2) **The decision was political in nature, even though it was a science-based issue that needed to be resolved.** While this relies on second- and third-hand sources, I believe the quote attributed to Ruckelshaus in a letter he wrote to American Farm Bureau President Allan Grant regarding the decision to ban DDT to be valid, in which he states: "In such decisions the ultimate judgement remains political."[63] While it can be argued that any decision made by a government official is "political," this decision was tainted from its association with environmental and activist groups that petitioned for the banning of DDT, fueled by the momentum of Carson's *Silent Spring*.

3) **Government action was taken when the free market was already resolving the issue.** This is actually a common argument in support of the decision to ban DDT, but I view it through a different lens. Global peak DDT usage occurred more than a decade before this decision was reached and usage continued to decrease every year after. As we learned more about pesticides and the desirable characteristics that should be screened for, like selectivity and low persistence, alternative pesticides were being discovered and registered that were replacing DDT in common agricultural uses. The abrupt discontinuation of the pesticide was unwarranted and caused undue stress on an industry that was already working to find better solutions.

4) **Downstream effects of the decision to ban DDT were not adequately considered.** I understand that DDT was not completely banned by this decision, and that it technically only applied to the United States, but the ramifications of the decision were felt across the globe. Exports of American-made DDT began to dry up and the costs for the pesticide rose to the point where poorer countries could no longer afford it. Even more troubling was the stance taken by the US Agency for International Development (USAID), establishing regulations that prohibited funding for any international projects that used DDT. The result was that developing countries were essentially forced to stop using the pesticide if they wished to receive the humanitarian aid they needed from the USAID, denying people a remedy that cost human lives. It has been estimated that over 100 million people have died in Africa from malaria that could have been prevented if not for banning DDT.

5) **The "environment" was prioritized over human life. Alternatively, adequate substitutes to DDT were not available prior to the cancelation of the pesticide.** In Ruckelshaus' decision to ban DDT, he stated that farmers would be provided instructions on how to use an alternative pesticide, methyl parathion. While this substitute is less persistent in the environment, it is significantly more toxic to humans. It is a factual statement to say that farmers were told to use a pesticide that posed a greater risk to their lives because the alternative was speculated to harm wildlife.

Would DDT stand a chance of being registered as a new insecticide today? Absolutely not! Residue and environmental fate studies would disqualify this product in early screenings before the effectiveness of the product could even be truly determined. As we will see in the next section, we have stringent mechanisms in place to ensure that any product that is registered meets the highest imaginable safety standards. However, the big question is: did the decision to ban DDT have unintended consequences on human life? Or put another way, was the right decision made? Whether intentional or not, my analysis of all available literature is that it was not. For support, we can look to the World Health Organization that in 2006 announced its support of DDT to once again quell malaria. According to Dr. Arata Kochi, director of WHO's global malaria program, "we must take a position based on the science and data. One of the best tools we have against malaria is indoor residual house spraying. Of the dozen insecticides WHO has approved as safe for house spraying, the most effective is DDT."[37]

It may seem like we took a *really* long rabbit trail going through the era of Rachel Carson and *Silent Spring*, but hopefully you can see how impactful that one book was on an entire industry, not to mention the health and safety of countless individuals from around the globe still dealing with insect-borne diseases like malaria. The review was also essential to understand the changing social and political landscape around the agricultural industry, particularly as we jump back into our regulatory review of pesticides.

— Regulations and the Agrochemical Industry . . . Again —

We could spend a lot more time on the subject of regulations in the agrochemical industry, but I don't want to put you to sleep with Congressional legislation! The goal is really to bring you into the complex world of agricultural chemicals and give you a glimpse into the layers of protection that have been put in place to ensure the public is protected from any potential poisons. With that said, the last piece of legislation we had covered was the Federal Food, Drug, and Cosmetic Act and the amendments to it in the 1950s. Now we move into the 1970s, into post-*Silent Spring* times, and an age where a new agency has been created, the Environmental Protection Agency (EPA). Regarding agricultural pesticides, a lot is about to transpire with amendments to FIFRA (the Federal Insecticide, Fungicide and Rodenticide Act . . . remember!).

The Federal Environmental Pesticide Control Act (FEPCA) of 1972 is up first. This act amended FIFRA by bringing the registration process for pesticides under the authority of the Administrator of the EPA. Additionally, pesticides would be classified as either "general" or "restricted" use. This created major changes in the industry because Restricted Use Products (RUPs) could now only be applied by, or under the direct supervision of, certified applicators. Today, many agricultural pesticides—whether herbicides, insecticides, fungicides, or others—are classified as Restricted Use. On the EPA's website, the most recent list of RUPs runs fifty-seven pages, which include 1,136 individual products.[38] I think it is important to highlight the continued efforts that are being made to ensure that those in the agricultural industry are both safe and mindful stewards of the technologies being created. Today, any individual that applies RUPs must be certified by the state (or states) in which they make applications and maintain their certification through continuing education credits.

The next amendment came a year later with the Endangered Species Act (ESA) in 1973. Remember when I was introducing Rachel Carson? This was one of the Acts that has been credited as stemming from the response to her book. This Act amended FIFRA to "include situations involving unreasonable hazard to the survival of a species declared by the Secretary of the Interior to be endangered or threatened." It also added more government agencies, like the US Fish and Wildlife Service (FWS) and the National Oceanic and Atmospheric Administration's (NOAA) National Marine Fisheries Service into the mix to work with the EPA.[39]

In the waning months of President Ronald Reagan's second term in 1988, another significant amendment was made to FIFRA, which, as Reagan stated, would "go a long way toward assuring safer pesticide use."[40] While there were changes to the funding requirements and indemnification clauses—all of which were absolutely critical, especially for a relatively new agency like the EPA that continued to struggle with limited budgets and funding—the main thrust of the amendment was dealing with the increasing number of older pesticides that required new evaluations for risk. This was an issue that was identified by chemical companies and environmental groups alike.

Tens of thousands of products had come to market since the start of WWII and the regulations we have already covered guaranteed there was a formalized process to register these products and ensure they were safe for humans, the environment, and beneficial organisms. However, that was also the problem. The system was only set up to test and register *new* products. As the industry evolved, so did its understanding of the potential risks of pesticides and its ability to test for these risks. Older pesticides on the market had not been tested with newer, more rigorous evaluations, and by the 1970s it was estimated that somewhere between 30,000 and 60,000 pesticides fit this description and needed *reregistration*.[41] Much of the 1970s and 1980s were dedicated to figuring out how to do this effectively, but the FIFRA Amendments of 1988 finally laid out the processes that would ensure all pesticides, regardless of when they were initially registered, had been evaluated with the latest methods and deemed safe for the public.

Alright, the moment you've all been waiting for. We've come to the last piece of legislation that we'll cover in this history lesson, jumping forward to just before the turn of the millennium during President Bill Clinton's time in office. The year was 1996. The piece of legislation was the Food Quality Protection Act (FQPA). I would argue that all the legislation covered so far focused on protecting the consumer or the health and safety of humans and the environment, but the FQPA arguably went the furthest to reduce potential risk to human health from pesticides. While this is beneficial towards my goal of providing peace of mind to those of you that believe organic food is the only safe option in the grocery stores, I would argue that FQPA actually went far beyond what is justifiable from a scientific standpoint. But let's keep reading to see what I'm talking about.

This legislation requires a bit more time to unpack because I think it will really open some eyes to the tolerances that have been put in place to safeguard our foods from pesticides. Tolerances, what does that word even mean? Since we are talking about chemicals and the impact on human health, we may want to take definition 4a(1) from Merriam-Webster that defines tolerance as "the capacity of the body to endure or become less responsive to a substance (such as a drug) or a physiological insult especially with repeated use or exposure."[42] Seems reasonable to me, especially as it references the "capacity of the body to endure" and "repeated use or exposure." Actually, this defines the way we used to think about tolerances established for pesticides relatively accurately. However, tolerance as it pertains to a pesticide has received its own definition on Merriam-Webster listed as entry 4b: "The maximum amount of a pesticide residue that may lawfully remain on or in food." It may not seem like a huge difference, but I would argue that one definition removes science from the equation completely, and that is the one that is currently used to regulate pesticides today.

So, what was new about the FQPA, and how did this change the agricultural pesticide industry? One was the introduction of "risk cup" into the lexicon of the agricultural sector. While the risk cup theory had been used prior to the new law, the guidelines introduced by the FQPA brought the concept to the forefront. The risk cup can be described as the total amount of a chemical to which a person can be safely exposed. When you add up all the potential exposures a person may have to a chemical, if it overflows the cup, it is now a potential risk. Clear as mud, right?

According to the EPA's webpage titled "Summary of the Food Quality Protection Act,"[43] they list the following four requirements that the act imposed:

1. Make a safety finding when setting tolerances, i.e., that the pesticide can be used with "a reasonable certainty of no harm;"

2. Use this new safety standard to reassess, over a 10-year period, all pesticide tolerances that were in place when the FQPA was signed;

3. Consider the special susceptibility of children to pesticides by using an additional tenfold (10X) safety factor when setting and reassessing tolerances unless adequate data are available to support a different factor;

4. Consider aggregate risk from exposure to a pesticide from multiple sources (food, water, residential and other non-occupational sources) when assessing tolerances; and consider cumulative exposure to pesticides that have common mechanisms of toxicity.

The last two points particularly affected the chemical industry because of the introduction of three concepts that would be added to the calculation of tolerances set for specific pesticides:

· **Safety factor** – An additional consideration for juvenile susceptibility

· **Aggregate risk** – The accounting for every possible route that may result in exposure to a pesticide

· **Cumulative exposure** – The accounting for pesticides with similar mechanisms of toxicity in the evaluation of any one of them.

Let's use an example to show how these new requirements changed the game.

Prior to the FQPA, a pesticide was evaluated based on its use pattern on food crops and its likely exposure to humans. If a pesticide was labeled for use on apples and pears, the maximum possible number of apples and pears that a human could consume per day, and by extrapolation, the maximum amount a human could consume in a lifetime if this level of consumption were maintained, was used to establish the potential exposure. Tolerances for pesticide levels were determined based on this theoretical potential exposure. It could be argued that this rationale was already providing a very conservative threshold for tolerances as the theoretical level of exposure would likely never be met by any individual. Oh, and by the way, the tolerance already had an "uncertainty factor" of 100-fold applied to it. This means that if toxicological data from studies show 1000 ppb of a pesticide to be safe, the tolerance is set to 10 ppb to provide a safety margin of error. So, the size of the risk cup is based on the tolerances set from toxicological data that has been reduced 100-fold.

Now let's look at the same example under the FQPA and how the new rules affected the risk cup. The first change was the addition of the *safety factor* that was established to protect the potential sensitivity of children to pesticides. This meant that if the toxicology studies did not evaluate children directly (and obviously most toxicology studies do not involve children as test subjects), the tolerance was automatically reduced by a factor of ten. So, if the tolerance was set at 10 ppb, which was already 100-fold less than the safe level established by the toxicology data (1,000 ppb in this case), it automatically was reduced to 10 ppb because of this new law. The risk cup was reduced in size tenfold from what it had been previously, or now 1,000-fold less than the level that data determined the product to be safe at.

Next up was the concept of *aggregate risk*. This requires that all potential routes of exposure be identified and added up to determine the potential exposure. The level of potential exposure calculated from the actual use on food crops was still used, but new requirements also considered potential run-off and contamination in drinking water, as well as any "residential or non-occupational sources." This means that if the pesticide was also used for lawn care, or pest control around the home, or even on golf courses or highway roadsides, it would also count towards the potential exposure to humans. Ultimately, the multiple uses would fill the risk cup more quickly than previous calculations that only considered direct exposure from food.

Finally, the *cumulative exposure* to pesticides that have common mechanisms of toxicity was required. Individual pesticides were no longer evaluated in isolation, but rather in combination with other pesticides that have similar modes of action—in other words, multiple pesticides now shared the same risk cup. This is a hard concept to describe, but I hope the following analogy I have formulated illustrates the importance of this new methodology.

In the world of over-the-counter (OTC) pain relievers, we, as consumers, have numerous options—Tylenol, Advil, Aspirin, Excedrin, etc. Fortunately, each of these drugs is evaluated for the dose required to be effective, while also being safe for human ingestion. Tylenol is limited to 4,000 mg/day, Advil to 1,200 mg/day, Aspirin to 4,000 mg/day, and so on. Now imagine that during the registration process, the regulators must assume that individuals will use all of the OTC options at the same time. Nevermind the fact that one may be used for headaches at one point in time, one may be used for joint pain at another point in time, and a third may be used for heart health benefits by a different individual.

Regardless of the probable use pattern, it must be assumed that all of these drugs will be taken at the same time, by the same individual. Therefore, the risk of exposure is set by establishing the tolerance of all drugs being taken together. Each individual drug can only constitute a proportion of the total exposure. The result would either be the reduction of the allowable dosage of each OTC drug, or the removal of certain drugs from the industry to allow for higher effective doses to be used by those that remain in the market.

This was the new burden placed on pesticides, and one of the industries that was most concerned during the initial years after this law passed was the home pest control industry. Because they were using many of the same pesticides for residential pest control (think ants, termites, and the like) as the agricultural industry was for insect pests of crops, something was likely going to have to give, and it probably wasn't going to be the more profitable, larger market of the agricultural industry. One article, titled "The Risk Cup: It Is Filling Fast" published to *Pest Control Technology* in 1998, reviewed this topic and provided feedback by members of the industry.[44]

Joe Yoder, Novartis' director of research and development for specialty products, said that "companies are going to defend the use that is defensible and economically viable. We may lose relatively small but critical uses for products." The thought was well stated; chemical companies would now need to look at all the potential uses of their pesticides and determine which uses could be eliminated to reduce the amount going into the risk cup. Another industry professional, Charles Frommer, president of Evins Pest Control, Long Island, NY, and active member of the National Pest Control Association's government affairs committee, remarked, "The smaller guys are the ones that are going to suffer the most: John PCO [pest control operator] and John Farmer. It's time to take a cold, hard look at the pesticides you're using now. Start looking for alternatives, so you're not caught at the last minute."

Okay, just to summarize. Tolerances had already been established to ensure the potential exposure of a pesticide through its application on food products was significantly less than the amount that could cause harm as determined by toxicology studies. To put it another way, the potential exposure couldn't realistically fill the risk cup. Now, however, more routes of exposure were added to the risk cup, more pesticides had to use the same risk cup, and the risk cup was reduced in size by a factor of ten automatically! The chances of the risk cup overflowing were much greater now and exposures that caused the risk cup to overflow would have to be eliminated. Every pesticide that was registered prior to the 1996 Act needed to be reassessed using these new guidelines to establish new tolerances. This means that every pesticide in use today has tolerances set based on these ultra-conservative guidelines.

The new law had a tremendous impact on the pesticide industry. The EPA reassessed 9,721 pesticide tolerances using the new guidelines in the ten-year timeframe that was required by the law, revoking or modifying the tolerances of nearly 4,000 of them.[45] There are critics of the EPA and the requirements set forth by the FQPA on both sides of the argument. Some do not think the regulations have gone far enough, especially regarding the protection of children. Some believe the cumulative and aggregate risk considerations are too conservative and provide undue scrutiny on exposures that are not going to have the same impact on humans as direct consumption. Either way, hopefully this detailed look at our current regulatory landscape provides you with an increased appreciation for the conservative processes and calculations that are used to ensure that we are safe from any adverse effects of pesticides.

Before we wrap up this section, let's review an example of how we view tolerances of pesticides differently than other chemicals to which we are routinely exposed. I came across an article by Dr. Alison Bernstein, professor of translational neuroscience at Michigan State University, that was posted to the Food and Farm Discussion Lab website.[46] It offers a brilliant comparison between two common chemicals, caffeine and glyphosate. We discussed glyphosate previously when we were looking at acute toxicities and stating that this was not a useful way to evaluate pesticide safety. But now that we have been speaking about toxicity studies, tolerances, thresholds, and exposure levels, I think we are ready to revisit this chemical and look at the relevant toxicological data. But first, a few terms that Bernstein uses, along with their definitions, are provided below to make sure we are all on the same page.

· **NOAEL (No Observed Adverse Effect Level)**: The *highest* chemical dose at which there *was not* an adverse effect.

· **LOAEL (Lowest Observed Adverse Effect Level)**: The *lowest* chemical dose at which there *was* an adverse effect.

· **RfD (Reference Dose)**: The daily amount of a chemical you can be exposed to every day for your entire life without resulting in an adverse effect.

· **Uncertainty factors (UF) and Multiplying factors (MF)**: Buffers used to reduce the RfD to provide more conservative values.

Okay, now that we have the acronym mumbo jumbo out of the way, why do I think this article is so brilliant? Because Bernstein provides data in a cogent way and draws parallels that are relevant for a useful discussion. She starts with caffeine, something most of us are extremely familiar with and deem necessary every morning to be a functioning, contributing member of society.

Bernstein cites a paper that summarized numerous prior studies (over 700 references) to show how the oral RfD of caffeine was established. Remember, the RfD is the "daily exposure" that humans can experience before being at risk of "deleterious effects throughout the entire lifetime." The paper, which surprisingly provided data for children in addition to healthy adults and pregnant women, showed a LOAEL for caffeine of 2.5 mg/kg/day. Again, this means that the lowest dose tested that resulted in adverse effects was 2.5 mg/kg/day. Three UFs, each providing a 10X conservative buffer, were then applied to the LOAEL value to 1) extrapolate from acute to chronic toxicity, 2) extrapolate from sub-chronic to chronic toxicity, and 3) account for variability in sensitivity within the population. All told, the three UFs added a 1,000X buffer to the LOAEL, resulting in an RfD of 0.0025 mg/kg/day.

$$RfD=LOAEL*UF1*UF2*UF3$$

$$RfD=(2.5\ mg/kg)/day*0.1*0.1*0.1$$

$$RfD=(0.0025\ mg/kg)/day$$

Bernstein uses herself as an example of how much caffeine this would equate to, but I will use our previous example from an earlier chapter looking at an average American male, aged 20-39, who would weigh 196.9lbs, or 89.3kg. Multiplying these numbers, we get a reference dose for caffeine of 0.223 mg/day. Bernstein provides the caffeine content of various coffee sources, but she states that a good rule of thumb is that an eight-ounce cup of coffee will contain 100 mg of caffeine. This means that, according to toxicology studies, an average American male is consuming 448 times the RfD of caffeine with just one cup of morning joe! Wow!

Next, Bernstein turns to glyphosate and again looks at a report that summarizes previous studies, this time from the European Food Safety Authority (EFSA) in the European Union. This summary has been well documented, and all research contained within is publicly available. In this instance, an overall long-term NOAEL for glyphosate is provided at 100 mg/kg/day for chronic exposures in animals. The EFSA uses a lower NOAEL for maternal and developmental toxicity of 50 mg/kg/day to calculate the RfD (the actual value in the study is called the acceptable daily intake (ADI), which is just the European terminology for the RfD). The total UF used in this case was 100-fold, resulting in a final RfD of 0.5 mg/kg/day. The EPA established its own RfD, which was even more conservative at 0.1 mg/kg/day. Taking the same average American male and using the more conservative EPA value, the reference dose of glyphosate is 8.93 mg/day. This means, that according to toxicology studies, the allowable daily exposure of glyphosate is 40 times higher than that of caffeine.

While we could just let the numbers speak for themselves and say "Point proved!" it is only fair to compare these values if we look at the exposure levels of each chemical. Just knowing that one is more toxic (caffeine, in this case) really does not tell us anything unless we know how much of each we are likely exposed to. This is the beauty of Bernstein's argument: she closes it out with the real comparison we should be looking at. She provides data from the Farm Family Exposure Study that looks at detectable levels of glyphosate in the urine of pesticide applicators, the population at highest risk for exposure to glyphosate. In the study, it was noted that 60% of applicators had detectable levels of glyphosate in their urine with the average concentration coming in at 3.2 ppb. The highest detectable level in the study was 223 ppb. Even if we take the highest reported level from the most at-risk population for exposure, the levels equate to 4% of the RfD, or 0.004 mg/kg/day (the average level was much lower, equivalent to 0.1% of the RfD).

Based on the toxicology data, caffeine is more toxic than glyphosate to humans. But even more relevant is that our exposure to caffeine is also much higher than it is to glyphosate. Bernstein concludes with this pointed statement regarding the public's perception of pesticides. "The exposure numbers above show that we don't give a second thought to consuming caffeine at levels hundreds of times higher than the oral RfD but are simultaneously worried about exposures to glyphosate that are 100 times lower than the RfD. In discussions of toxicity, we must use the correct data to back up our points to step outside the cycle of misinformation."

When we take the time to evaluate the regulations that have been put in place to keep us safe, it sheds light on how risk-averse our governing bodies are when it comes to the adverse effects of pesticides. There was clearly a lot of work to be done when the agricultural pesticide industry first got into full swing in the 1940s and 1950s, but through the collaborative work of agricultural groups, environmental groups, and the government, we have made tremendous strides. In my opinion, we should feel safe about eating foods produced by any practice in the US, whether organic or conventional.

— I Heard Roundup Causes Cancer —

This will be a relatively short section, but given that we dedicated significant time to the toxicity of glyphosate, I thought we should at least cover the topic of glyphosate causing cancer. These days, it is nearly impossible to turn on the TV without seeing an ad for some law office telling you to call in if you or a family member have suffered from non-Hodgkin lymphoma because it could be linked to glyphosate and you "could be entitled to a large settlement." We need to put this to bed.

Where did this all stem from anyway? In March 2015, the International Agency for Research on Cancer (IARC), a subsidiary agency of the World Health Organization, released a monograph classifying glyphosate as a Class 2A chemical, or one that is "probably carcinogenic to humans."[47] Well, that sounds bad! Let's take a quick step back and understand what the IARC is and what its classification system is all about.

The IARC was created in 1965 to "lighten humanity's ever-growing burden of cancer," according to their website. Based in France, the organization has grown from the original five charter countries (France, Germany, Italy, the United Kingdom, and the United States) to twenty-six countries from around the world. The IARC's mission is to "coordinate international studies on the causes of human cancer, the mechanisms of carcinogenesis and strategies for cancer prevention, with a particular focus on promoting research in regions of the world where it is lacking."[48] While it has the stated goal of producing original research, the IARC has received the most attention from its carcinogenicity classification system for a wide range of products and activities.

Soon after the creation of the organization, the IARC suggested a "compendium on carcinogenic chemicals be prepared by experts. The biological activity and evaluation of practical importance to public health should be referenced and documented." The program was approved in 1971 and the IARC has been classifying chemicals, and even activities, as cancer hazards ever since.

It was ever so subtle in that last sentence, but I used the word *hazard* instead of *risk*. This was intentional and particularly important to the discussion at hand. The word hazard has only been used four times in this book so far, and each instance was in a quotation from a law or amendment describing either DDT or the protection of wildlife. In every other discussion we have had around chemicals and pesticides, we have used the word risk. Why does this matter?

Instead of me answering this question for you, let's have the IARC, as written in the preamble for their classification monographs, explain it themselves: "A cancer hazard is an agent that is capable of causing cancer, whereas a cancer risk is an estimate of the probability that cancer will occur given some level of exposure to a cancer hazard... The Monographs identify cancer hazards even when risks appear to be low in some exposure scenarios."[49] Huh, that's interesting. So, all the discussion we have had around the study of toxicology and how the dose makes the poison—or how the science involved in pesticide safety is continuing to evolve and look at different pathways of exposure—all that is for naught according to the IARC.

Let's use a few analogies to clarify this even further. A body of water is a hazard, but risk comes into the equation when a person jumps into that body of water without knowing how to swim. Or consider gasoline; that is certainly a hazard. Gasoline has the potential to cause harm, but there is relatively little risk associated with an open can of gasoline until you light a match next to it. If we went through life making decisions, consciously or unconsciously, based on hazard assessments instead of risk assessments, we would never leave the house. Now that I think about it, our house is full of hazards too! Electricity, bathtubs, food, water, glass, pets . . . Are you starting to freak out a little bit?

But this is the approach IARC takes to classify chemicals and situations as cancer "agents," and this is why they are still the lone scientific body that has come to the conclusion that glyphosate is "probably carcinogenic to humans." An infographic from the Genetic Literacy Project lists the statements made by other regulatory and research agencies on glyphosate, of which a few examples are provided below:[50]

"Unlikely to be carcinogenic to humans or genotoxic (damaging to genetic material or DNA) and should not be classified as a mutagen or carcinogen."
– Environmental Protection Authority (New Zealand, 2016)

"Glyphosate is unlikely to be genotoxic or to pose a carcinogenic threat to humans . . . Neither the epidemiological data nor the evidence from animal studies demonstrates causality between exposure to glyphosate and the development of cancer in humans."
– European Food Safety Authority (European Union, 2015)

"No pesticide regulatory authority in the world currently considers glyphosate to be a cancer risk to humans at the levels at which humans are currently exposed."
– Health Canada (Canada, 2019)

"Human health risk assessment concludes that glyphosate is not likely to be carcinogenic to humans . . . [and] no other meaningful risks to human health when the product is used according to the pesticide label."
– US Environmental Protection Agency (United States, 2017)

Even the World Health Organization, the governing body of the IARC, stated in 2016 after the IARC released the new classification:

"Glyphosate is unlikely to be genotoxic at anticipated dietary exposures. Glyphosate is unlikely to pose a carcinogenic risk to humans from exposure through diet."

Additionally, we have the benefit of being able to conduct long-term studies to evaluate the effect of glyphosate on humans since it was first produced in 1971. The Agricultural Health Study, funded by the National Cancer Institute and National Institute of Environmental Health Sciences, continually monitors a "cohort" of pesticide applicators. The aim of the research project is to "understand how agricultural, lifestyle, and genetic factors affect the health of the farming population."[51] They have monitored and evaluated the cohort of 54,251 pesticide applicators since 1993. Not a bad sample size for any of my statisticians in the audience. And their finding . . .

"No association was apparent between glyphosate and any solid tumors or lymphoid malignancies overall, including non-Hodgkin lymphoma and its subtypes."
– Agricultural Health Study, United States, 2018)

I know science is not governed by consensus, but the overwhelming body of data and opinions of experts from around the world should lead the rational observer to the same conclusion: glyphosate is not likely to cause cancer. But, we have the lone voice of the IARC to contend with:

"There is limited evidence in humans for the carcinogenicity of glyphosate. A positive association has been observed for non-Hodgkin lymphoma . . . There is sufficient evidence in experimental animals for the carcinogenicity of glyphosate . . . Glyphosate is probably carcinogenic to humans (Group 2A)."
– International Agency for Research on Cancer (2015)

Well, I guess it's settled then.

It is important to spend time on this subject because the ramifications of the IARC study have been far-reaching. As previously mentioned, you can hardly turn on the TV anymore without hearing an ad for a law firm talking about the opportunity to be part of a class-action lawsuit against Monsanto (now Bayer). And the commercials are being put out there for good reason. There have now been three successful lawsuits brought against Monsanto, all in California courts, where the juries have ruled in favor of the plaintiffs. Although there is no evidence that the chemical has ever caused cancer in humans, or that a cancer risk even exists, juries were somehow able to reach the verdict that a company must pay individuals $289.2 million in *Johnson v Monsanto* (reduced by the judge to $78.5 million), $80 million in *Hardeman v Monsanto*, and **$2.055 billion** in *Pilliod v Monsanto*.[52] With tens of thousands more potential plaintiffs waiting in the wings for their payday, Bayer AG, who purchased Monsanto in 2018, reached an agreement in 2020 to settle any outstanding lawsuits and those to come in the future to a tune of $10 billion. That's *billion*, with twelve zeroes!

I guess the next question we should ask is when the lawsuits are going to start coming for Starbucks, Texas Roadhouse, Super Cuts, and maybe even Whole Foods? You may be thinking, "That's a random segue," but not really. The lawsuits came because the IARC classified glyphosate as a group 2A agent, which means "probably carcinogenic to humans." Other "agents" that fall into this same classification are coffee, red meat, working as a hairdresser, and night shift work. So, yes, Starbucks, Texas Roadhouse, Super Cuts, and even Whole Foods (which has the audacity to employ people to work the night shift), I'd get ready if I were you.

We have focused primarily on group 2A from the IARC, but they currently have three other classifications that agents can fall into: group 1 (carcinogenic to humans), group 2B (possibly carcinogenic to humans), and group 3 (not classifiable to its carcinogenicity to humans). Prior to 2019, there was a fifth category, group 4 (probably not carcinogenic to humans). However, of the 1,013 agents the IARC had reviewed, only one had been found to fall into this last group, which was the chemical caprolactam, a precursor of nylon, fibers, and plastics. So, the IARC decided to remove this category altogether and state that they could not claim that any agent was "not carcinogenic." A study published in *Toxicology Research and Application* pulled a random sample of 100 agents classified as group 3 (not classifiable) by the IARC and found that 24% had no structural elements that would enable them to cause cancer. [53] The authors conclude that there is a "reluctance on the part of IARC to place agents into the lowest category of risk." Contrarily, we see that there is no such reluctance to place agents in higher categories of risk, even when the overwhelming data would suggest otherwise.

After about 80 pages together, I feel like we have reached a point where we can speak openly with each other. My overall contention with all of this is that the regulation and marketing of agriculture and food production have become very polarized. And, if I am honest with you and myself, I am part of the problem as I find myself defending certain practices or looking down at others simply because they are promoting a different "methodology" than I would use myself.

It seems as though public sentiment carries more weight than scientific findings, and this is why I decided to write a book: to challenge myself to look at the arguments and hot topics surrounding the industry that I care so deeply about and ensure that I have sufficiently investigated the available data. It was only after weeks of diving into this challenge that I realized several pages had miraculously materialized and it may be worth sharing my thoughts and findings with others that may not have the time to investigate these topics themselves. Wherever polarization occurs, we must meet it head-on and turn towards fact-based reasoning.

Let's look at one example of this polarization before we move on. Glyphosate is arguably under the most scrutiny in the European Union. In 2018, the EU authorized the use of glyphosate through December 2022, but it is still a hotly contested topic. The EFSA has concluded that the herbicide poses no risk to humans, but the public outcry that followed the IARC finding is driving policy.

This is in stark contrast to copper sulfate, a fungicide that was previously mentioned as an approved pesticide for disease control on the National List of Allowed and Prohibited Substances for organic food production. Unlike glyphosate, the EFSA has found that copper compounds, like copper sulfate, are "of particular concern to public health and the environment."[54] Other European agencies have reported similar warnings about the risk of copper sulfate, like the French National Institute for Agricultural Research (INRA) that was co-commissioned by the French Institute for Organic Farming (ITAB). The INRA found that excessive levels of copper sulfate can "have adverse effects on the growth and development of most plants, microbial communities, and soil fauna" and recommend the government to "reduce use of copper for the protection of biological uses."

Obviously, if the EU is taking such a hard stand against a chemical like glyphosate, an herbicide that they have found to pose no risk to humans or wildlife, surely they would be adamantly opposed to the use of a pesticide that they have concluded to be a risk, right? This is where the hypocrisy is on full display.

Organic groups are coming out in force to show their support for the continued use of copper sulfate to control diseases in organic grapes, all while jeering the efforts of farmers to maintain the use of the proven, safe herbicide glyphosate to control weeds in their fields.

The EU political group Progressive Alliance of Socialists and Democrats falls into this camp. The group's chief of the committee that monitors the transparency of pesticide authorization in the EU, Eric Andrieu, explicitly stated that public health should take precedence over economic interests, but changed his tune when it came to copper sulfate. In this case, he states "alternatives to copper remain very limited and currently do not meet the demand of 500 million consumers. In the short term, the survival of a large part of European winery, in particular, the organic winery is at stake."[55] This may be a low blow, but I don't remember the same leniency afforded to DDT when alternatives were "limited." And it wasn't the organic wine industry at stake in that instance, it was human lives across much of the globe! What's good for the goose is good for the gander, as they say, unless the goose is a non-organic herbicide that has been proven safe by numerous scientific bodies from around the world and the gander is an approved organic fungicide that has been shown to be harmful to public health and the environment.

1 Natural Grocers (2017). The top 3 reasons shoppers buy organic produce. PR Newswire. https://www.prnewswire.com/news-releases/the-top-3-reasons-shoppers-buy-organic-produce-300503419.html

2 Borzelleca JF (2000). Paracelsus: Herald of modern toxicology. *Toxicological Sciences: an official journal of the Society of Toxicology 53*(1), 2-4. https://doi.org/10.1093/toxsci/53.1.2

3 Melina R (2010). Why do medical researchers use mice? Live Science. https://www.livescience.com/32860-why-do-medical-researchers-use-mice.html

4 UC Regents (2004). Lethal dose table. LHS Living by Chemistry. https://whs.rocklinusd.org/documents/Science/Lethal_Dose_Table.pdf

5 Roland J and Biggers A (2019). *What's the average weight for men?* Healthline. https://www.healthline.com/health/mens-health/average-weight-for-men

6 The Extension Toxicology Network (1996). EXTOXNET PIP - Glyphosate. http://extoxnet.orst.edu/pips/glyphosa.htm.

7 Dagan M (2020). History of malaria and its treatment. In Patrick GL (ed.), *Antimalarial agents: Design and mechanism of action* (1-48). Elsevier.

8 Berry-Caban CS (2011). DDT and Silent Spring: Fifty years after. *Journal of Military and Veterans' Health* 19(4). https://jmvh.org/article/ddt-and-silent-spring-fifty-years-after/

9 Conlon JM: *The historical impact of epidemic typhus.* http://entomology.montana.edu/historybug/typhus-conlon.pdf

10 Dustman R (2013). *WWII military health in the Pacific.* AAPC. https://www.aapc.com/blog/26557-wwii-military-health-in-the-pacific/

11 Paltzer S (2021). *The other foe: The U.S Army's fight against malaria in the Pacific Theater,* 1942-45. Army Historical Foundation. https://armyhistory.org/the-other-foe-the-u-s-armys-fight-against-malaria-in-the-pacific-theater-1942-45/

12 Ibid.

13 Berry-Caban CS, DDT and Silent Spring

14 Ganzel B: *Farming in the 1940s: Pesticide regulations – FIFRA.* Wessels Living History Farm. https://livinghistoryfarm.org/farminginthe40s/pests_07.html

15 RachelCarson.org (2021). *Silent Spring.* The Life and Legacy of Rachel Carson. http://www.rachelcarson.org/SilentSpring.aspx

16 Brinkley D (2012). *Rachel Carson and JFK, an environmental tag team.* Audubon. https://www.audubon.org/magazine/may-june-2012/rachel-carson-and-jfk-environmental-tag-team

17 Papers of John F. Kennedy (1963). *Use of pesticides: A report of the President's Science Advisory Committee.* Presidential Papers. President's Office Files. Departments and Agencies. https://www.jfklibrary.org/asset-viewer/archives/JFKPOF/087/JFKPOF-087-003

18 Brinkley D (2012). *Rachel Carson and JFK.*

19 Dusquesne University (2021). *Charles Rubin.* Duquesne University Faculty. https://www.duq.edu/academics/faculty/charles-rubin

20 Rubin CT (2012). *Reading Rachel Carson.* The New Atlantis. https://www.thenewatlantis.com/publications/reading-rachel-carson

21 Burnet M (1958). Leukemia as a problem of preventative medicine. *New England Journal of Medicine* 259:423-31. https://www.thenewatlantis.com/publications/reading-rachel-carson

22 Wikipedia contributors (2021). Robert Zubrin. In *Wikipedia: The Free Encyclopedia*. https://en.wikipedia.org/wiki/Robert_Zubrin

23 National Audubon Society (2020). *The Christmas bird count historical results.* https://netapp.audubon.org/cbcobservation/

24 Grier JW (1982). Ban of DDT and subsequent recovery of reproduction in bald eagles. *Science 218*(4578):1232-5. doi: 10.1126/science.7146905. PMID: 7146905.

25 Wurster CF (1968). DDT reduces photosynthesis by marine phytoplankton. *Science 159*:1474-1475. https://doi.org/10.1126/science.159.3822.1474

26 Ehrlich PR (1968). *Ecocatastrophe* (24-28). City Lights Books.

27 Jukes TH (1971). DDT, human health and the environment. *Boston College Environmental Affairs Law Review 1*(3), 534. https://lawdigitalcommons.bc.edu/ealr/vol1/iss3/4

28 Wikipedia contributors (2021). The population bomb. In *Wikipedia, The Free Encyclopedia*. https://en.wikipedia.org/wiki/The_Population_Bomb

29 The World Food Prize Foundation (2021). *About Norman Borlaug.* World Food Prize Foundation. https://www.worldfoodprize.org/en/dr_norman_e_borlaug/about_norman_borlaug/

30 Skorup J (2012). *The greatest man you've never heard of: Norman Borlaug, an American hero.* Michigan Capitol Confidential. https://www.michigancapitolconfidential.com/17495

31 Cremer J (2020). *Norman Borlaug saved millions of lives, would his critics prefer he hadn't?* Cornell Alliance for Science. https://allianceforscience.cornell.edu/blog/2020/04/norman-borlaug-legacy-documentary/

32 Wikimedia Commons contributors. File: J. Gordon Edwards eating DDT.jpg. *Wikimedia Commons, The Free Media Repository.* https://commons.wikimedia.org/wiki/File:J._Gordon_Edwards_eating_DDT.jpg

33 Walker J (2004). *J. Gordon Edwards.* The Blog of Death. http://www.blogofdeath.com/2004/07/26/j-gordon-edwards/

34 Edwards JG (2004). DDT: A case study in scientific fraud. *Journal of American Physicians and Surgeons 9*(3). https://www.jpands.org/vol9no3/edwards.pdf

35 Edwards JG (1992). The lies of Rachel Carson. *21st Century Science & Technology Magazine.* https://21sci-tech.com/articles/summ02/Carson.html

36 *Consolidated DDT Hearing: Hearing Examiner's Recommended Findings, Conclusions, and Orders, EPA*, 40 C.F.R. § 164.32 (1972) (examiner Judge Edmund Sweeney). https://www.thenewatlantis.com/wp-content/uploads/legacy-pdfs/20120926_SweeneyDDTdecision.pdf

37 Rehwagen C (2006). WHO recommends DDT to control malaria. *BMJ 333*(7569). https://doi.org/10.1136/bmj.333.7569.622-b

38 Environmental Protection Agency (2019). *Restricted use product summary report.* https://www.epa.gov/sites/production/files/2019-10/documents/rup-report-oct2019.pdf

39 US FWS: *Federal Environmental Pesticide Control Act of 1972* (7 U.S.C. § 136). Digest of Federal Resource Laws of Interest to the U.S. Fish and Wildlife Service. https://www.fws.gov/laws/lawsdigest/FEDENVP.HTML

40 US EPA Press Office (1988). *EPA history: FIFRA amendments of 1988*. United States Environmental Protection Agency. https://archive.epa.gov/epa/aboutepa/epa-history-fifra-amendments-1988.html

41 Ferguson S and Gray E (1989). *1988 FIFRA amendments: A major step in pesticide regulation*. The Environmental Law Reporter, 19 ELR 10070. https://elr.info/sites/default/files/articles/19.10070.htm

42 Merriam-Webster (n.d.). Tolerance. In *Merriam-Webster.com Dictionary*. https://www.merriam-webster.com/dictionary/tolerance

43 US EPA (1996). *Summary of the Food Quality Protection Act: Public law 104-170*. United States Environmental Protection Agency: Laws and Regulations. https://www.epa.gov/laws-regulations/summary-food-quality-protection-act

44 PCTOnline (1998). *The risk cup: It is filling fast*. Pest Control Technology Online. https://www.pctonline.com/article/the-risk-cup--it-is-filling-fast/

45 US EPA, Summary of the Food Quality Protection Act

46 Bernstein A (2017). *Glyphosate vs. caffeine: Acute and chronic toxicity assessments explained*. Food and Farm Discussion Lab. http://fafdl.org/blog/2017/04/13/glyphosate-vs-caffeine-acute-and-chronic-toxicity-assessments-explained/

47 IARC Working Group (2017). Some organophosphate insecticides and herbicides. *IARC Monographs on the Evaluation of Carcinogenic Risks to Humans 112*. International Agency for Research on Cancer, World Health Organization.

48 IARC (2020). *IARC – A unique agency: Cancer research for cancer prevention*. International Agency for Research on Cancer, World Health Organization. https://www.iarc.who.int/wp-content/uploads/2020/06/IARC-brochure-EN-June_2020-web.pdf

49 IARC (2019). Preamble. *IARC Monographs on the Identification of Carcinogenic Hazards to Humans*. International Agency for Research on Cancer, World Health Organization. https://monographs.iarc.who.int/wp-content/uploads/2019/07/Preamble-2019.pdf

50 Schreiber K (2015). *What do global regulatory and research agencies conclude about the health impact of glyphosate?* [Infographic] Genetic Literacy Project. https://gmo.geneticliteracyproject.org/wp-content/uploads/2015/12/glyphosateinfographic-glp-.png

51 NIH (2020). *Agricultural Health Study*. National Institutes of Health. https://aghealth.nih.gov/

52 Goldman RLM and Moore DM (2021). *Monsanto Roundup settlement*. Baum Hedlund Law. https://www.baumhedlundlaw.com/toxic-tort-law/monsanto-roundup-lawsuit/monsanto-roundup-settlement/

53 Smith CJ and Perfetti TA (2019). An approximated one-quarter of IARC Group 3 (unclassifiable) chemicals fit more appropriately into IARC Group 4 (probably not carcinogenic). *Toxicology Research and Application*. https://doi.org/10.1177/2397847319840645

54 Porterfield A (2020). Pesticide hypocrisy? EU edges toward banning glyphosate after finding it safe but clears organic copper sulfate after finding it a 'public health and environment concern.' Genetic Literacy Project. https://geneticliteracyproject.org/2020/09/04/examining-the-eus-contradictory-treatment-of-glyphosate-and-copper-sulfate-pesticides/

55 Ibid.

Organic Food Production: The Verdict On Organics

— It's Organic . . . So it Has to Be Better for Me —

To close out our conversation on organic food production versus conventional practices, let's dive into the idea that organic food and farming are healthier for you, better for society, and more environmentally friendly. I mean, why else would you pay more for the same food if it were not a healthier option—or to a lesser degree aligned with an ideology around social and environmental consciousness? I would argue that organic foods are not an improvement over their conventional counterparts in these arenas, and are more likely inferior. But instead of inflammatory statements, let's review some facts so you can make up your own mind.

In reviewing scientific articles that evaluate the relative health benefits of organic and conventional food, it is hard to find a study that definitively states that organic food has an edge. Take for instance a 2012 piece in the *Annals of Internal Medicine* titled "Are organic foods safer or healthier than conventional alternatives."[1] The authors of this paper provide a comprehensive review of seventeen human studies and 223 food nutrient and contaminant level studies, but ultimately determine that the current body of work on the subject "lacks strong evidence that organic foods are significantly more nutritious than conventional foods."

In another, more recent, review published in *Critical Reviews in Food Science and Nutrition* titled "Organic food and the impact on human health," the authors were able to identify studies that noted measurable differences between organic and conventional food.[2] The finding that leaned most in favor of organic foods was the presence of bioactive compounds. These compounds, according to the article, are "healthy components such as polyphenols . . . [that] are secreted by plants in response to stress stimuli." As a plant pathologist, I find this to be a fascinating rationale for the use of organic farming practices. Essentially, the contention of the authors is that plants produced in organic farming systems are more stressed because of poorer nutrition, increased disease and insect pressure, and more detrimental conditions, so they may produce beneficial compounds as a defense mechanism. Not necessarily the way you want to approach raising crops if you are striving to increase yields and feed more people!

Perhaps a more important revelation from authors was the fact that no studies have shown that the increase in bioactive compounds actually results in a benefit to humans. No differences in antioxidant levels. No differences in the bioavailability of the polyphenols and other bioactive compounds. Even though the "healthy components" are present because the plants had a harder life, it appears that we don't reap any of the benefits. One difference that *is* observed and *has* a definite impact on us, according to the authors, is the increased frequency of microbiological contaminations in organic food. Both *Escherichia coli* and mycotoxin contamination occur more often in organic food than conventional food.

From my point of view, the most important point highlighted by the authors is found in the second paragraph of their conclusion. They state, "The health outcomes reported by some studies could also be closely linked to the lifestyle of organic food consumers." There it is. This exemplifies the common thread found in many studies and articles on this topic. It is legitimately difficult to find a true research study that can provide any concrete evidence to support the idea that organic foods are more nutritious or beneficial than conventional foods. Rather, study participants that consume organic food are often healthier because organic food selection is only one of many lifestyle choices in which this cohort tends to partake.

A study in the *European Journal of Nutrition and Food Safety* further proves this point by illustrating the positive lifestyle characteristics that organic food consumers choose.[3] Organic food buyers consumed more fruit (+17%) and vegetables (+23%) than their non-organic food buyer counterparts while consuming 58% fewer soft drinks and having more positive responses related to smoking, physical activity, and body weight. We can all agree that these characteristics are good, but they are not a result of the inherent quality or nutritional superiority of organic foods.

The reality today is that many people, even those in academia who should be objective, have an inherent and subjective belief that organic food is somehow more natural, so therefore it must be better for you. It is common to see a throwaway line in papers and articles about organic products having less pesticide residue, which often is true. And yes, this is frequently followed by a generic statement that pesticides *can* cause deleterious effects in humans. But this ignores the entire premise of toxicity, risk, and the tolerances established for pesticides that we have taken great pains to unpack over the previous sections. Instead, advocates for organic food are putting a belief ahead of the available facts, and as a result, a narrative is delivered without the supporting evidence.

When the national organic standards were created in 2000, the Secretary of Agriculture was Dan Glickman, an appointee of President Bill Clinton. Referring to the legislation that was passed to establish the National Organic Program, Glickman stated, "Let me be clear about one thing, the organic label is a marketing tool. It is not a statement about food safety. Nor is 'organic' a value judgment about nutrition or quality."[4] I think that is the real takeaway from the few studies we have looked at in this section. There is a belief that organic foods should be better because we have bought into the marketing tool that is the USDA Organic certification.

A positive consequence of this marketing tool could be the beneficial drive to implement healthy lifestyle choices, as is seen in research studies like those we have reviewed. However, it does not mean, from an empirical standpoint, that organic foods are better. And what about those consumers who have bought into the organic marketing machine and believe they need to purchase organic foods for the safety of their families? What about those consumers who don't have the expendable income to pay more for the same food products but feel an obligation to their loved ones? What are the unintended consequences in these scenarios?

— But It's Better for the Soil —

While health and safety concerns are the primary reasons cited for choosing to purchase organic foods, another common driver is the notion that organic food production is in some way better for the environment. Again, it is the ideal that organic agriculture is in some way more "natural" that has been promulgated. While it is understandable that the perception of organic farming, portrayed as some sort of throwback reminiscent of a Norman Rockwell painting, conveys an emotion that it is somehow better for the environment, the data simply does not support the ideal. Let's go through some of the arguments used to justify the "environmentally-friendly" status of organic agriculture—and also some of the glaring facts that cannot be overlooked.

When we think about farming, it all starts with the soil, and this tends to be a big focus of organic farming advocates. If farmers were painters, the soil would be their canvas. It is at the core of every farming operation and, quite frankly, it is one of the farmer's most important investments. And just like any good investor, a farmer wants to protect their investment and reap the returns. This goes for both organic *and* conventional farmers.

As we've discussed previously, there is no reason to impute nefarious intentions on another individual simply because they choose to conduct their business in a different way than your own. But if you were to simply go off the narrative driven by organic activists, you would assume that the soil is nothing more than a commodity for conventional farmers to exploit for maximum short-term gain, regardless of the long-term consequences. I have worked with conventional farmers directly for the past sixteen years, and I can tell you with certainty, this is not the case. But let's look at the arguments in front of us.

A statement that I often hear from organic supporters is that organic farming is better for soil quality and soil health. Besides these being very nebulous terms, I don't think that the underlying sentiment is valid. But that is the argument, and not just from fringe activists. We can start by looking at the Food and

Agriculture Organization (FAO) of the United Nations (UN) and what they say on the matter. According to the Organic Agriculture FAQ section of the FAO webpage, organic agriculture improves the structure and formation of the soil, reduces soil erosion, increases the retention of water and nutrients, increases biodiversity, and "enhance[s] soil productivity."[5] Well, sign me up!

So, should we take the word of a global organization like the UN at face value? How about the Organic Trade Association (OTA)? Obviously, this group is going to be biased towards promoting the practice that they were founded to advocate on behalf of, but what do they say on the matter? Brace yourself, it sounds a lot like the FAO. According to their website on a page titled "Environmental Benefits of Organic," they state that "instead of relying on synthetic pesticides and fertilizers that can deplete the soil of valuable nutrients and increase environmental degradation, organic agriculture builds up soil using such practices as composting, cover cropping, and crop rotation."[6]

I must confess that most of what the organic advocates have to say on this topic is absolutely correct. You heard me! Man, it feels good to get that off my chest. I concede that 90% percent of the FAO and OTA views on the environmental benefits of organic agriculture are correct. Improved soil characteristics, better nutrient retention, the focus on crop rotation and cover cropping, all accurate. Really, there is only one inaccuracy that I can point to—and that is the implication that organic agriculture is the *only* solution that delivers these benefits.

The fundamental problem we are up against with this whole debate is that it has become an either/or decision. Because organic farming has become an ideology, people are coerced into thinking that the dogma of organic farming is wholly unique—and that it is the only way to save the planet. But that is simply not the case. The organic farming movement has brought forth numerous positive ideas and practices that are being implemented by conventional farmers. As I mentioned before, the soil is arguably a farmer's most important asset, and they want to ensure they are the best stewards of it they can be.

The OTA stated that organic agriculture "builds up" the soil through cover cropping and crop rotation. Cover crops are a relatively new idea in "mainstream" agriculture, but one that is picking up steam across much of the country. If you are not familiar with this practice, the USDA provides a very comprehensive definition that is presented below:

"Crops including grasses, legumes and forbs for seasonal cover and other conservation purposes. Cover crops are primarily used for erosion control, soil health improvement, weed and other pest control, habitat for beneficial organisms, improved water efficiency, nutrient cycling, and water quality improvement...The cover crop may be terminated by natural causes such as frost, or intentionally terminated through management such as chemical application, crimping, rolling, tillage, grazing, or cutting."[7]

There are numerous benefits of cover crops stated in that definition, but these benefits will often vary depending on the type of cover crop that is planted. But the one thing all cover crops have in common is that they are not harvested. Because the cover crop is not meant to be harvested, there is no immediate value from sowing the crop and then cutting it, plowing it, or letting it die before the next crop is planted. Essentially, a farmer must believe that there will be longer-term economic or environmental returns because cover cropping is an expense from which they will not reap an immediate profit.

I think many of us would admit that investing for long-term gains, especially in non-traditional methods, in lieu of shorter-term, more conventional investments would not be an easy proposition to swallow. But as the research has accumulated on the subject, more farmers are experimenting with the idea. Yes, even conventional farmers. In fact, cover crop acres grew 50% on conventional farms between 2012 and 2017 to a total of 15.4 million acres.[8] While this still only accounts for 5.1% of all harvested cropland, it is increasing rapidly as more is learned about the practice. To put it in perspective, the 15.4 million acres of cover crops planted in conventional agriculture systems equate to roughly 4.5 times the total number of acres in organic cropland, which clocked in at 3.5 million acres in 2019.[9]

And what about the organic farmers? If this is a pillar of organic farming that studies and organizations cite as a major benefit, they must be planting cover crops on nearly every acre, right? Wrong. According to the USDA's 2019 Organic Survey, the percent of organic farms that planted cover crops was 31.9%.[10] To reiterate, that is the number of farms, not acres. In conventional farming operations, it was found that smaller operations were much more likely to plant cover crops than larger operations. In fact, 70% of the farms that planted cover crops were less than fifty acres.[11] Because this data does not exist for organic farming operations, we are left to speculate. However, if the trend of smaller conventional farms adopting cover crops holds true for organic farms, the percentage of organic *acres* that receive a cover crop is likely much less than 30%. Interesting, huh?

food for thought...

Precocious puberty. While you may have never heard those two words strung together in that order, you are likely aware of the condition they describe—the early onset of puberty. The reason this condition gets its own Food for Thought is because it was originally thought (and still is in a large portion of the population) to be caused by hormones in our food, specifically those found in milk. What is the backstory and what do we know now?

According to the Mayo Clinic, puberty is defined as "early onset" when it occurs before the age of eight in girls and nine in boys.[4] The condition is not new, but it was often classified as a "monstrosity" in early modern times. One of the first documented cases was that of Hannah Taylor from England in 1695, "a very extraordinary child of about six years of age, who in face, etc., was as large as a full grown woman."[12]

From the late 1800s to the mid-1900s, health records in Europe and the United States showed that the average age of puberty onset was trending earlier. Although the trend appeared to plateau for a few decades, the issue gained awareness in the 1990s. Anecdotal accounts from pediatric physicians began noting that puberty...

food for thought... (cont'd)

seemed to be, once again, occurring earlier than what was expected, particularly with breast and pubic hair development in young girls.[13]

In 1997, the results of a large study that evaluated 17,000 girls across the United States reported that initial signs of puberty were, in fact, occurring earlier than expected.[14] They also noted significant differences between different racial backgrounds. While there were other conflicting studies on the average age of puberty onset, the public was now aware that their children may be entering adolescence earlier than originally expected. The only thing missing was an explanation of why, and by perhaps nothing more than bad timing, hormones in milk became the scapegoat.

The bovine growth hormone (bGH), also referred to as bovine somatotropin (bST), regulates milk production in dairy cows and is found in all cow's milk. The biotechnology company Genentech, founded by Herbert Boyer (who we will refer to in the non-GMO section) discovered the gene for bST and synthesized the hormone artificially through recombinant technology (rbGH or rbST). In 1994, Monsanto commercially launched an injectable version of rbST to increase milk production in dairy cows under the trade name Posilac. The product was eventually sold to Elanco, and most recently to Brazilian-based Union Agener in 2018. Through the multiple owners of the product, one thing has remained constant—increased milk production. A marketing page for Posilac highlights this by stating "a healthy cow supplemented with Posilac produces an average of 10 more pounds of milk per day."[15]

The rbST active ingredient went through the complete FDA review process required for registration in 1993 to assess its safety for humans. The results were conclusive: "bST is degraded by digestive enzymes in the gastrointestinal tract and . . . [has] no biological activity . . . bST does not promote biological activity in the human body because somatotropins from lower mammalian species have no activity in humans."[16] But, timing is everything. The launch of a "hormone" during a time when puberty was thought to be occurring earlier was seen as more than coincidence.

I would classify the response to rbST as "premature outrage," because exposés...

that drew attention to a "new hormone" in our milk drove an emotional response without any scientific backing. It was even featured on an episode of The Simpsons as late as January 2016, titled "Teenage Mutant Milk-Caused Hurdles," where Bart sprouts a mustache, Lisa develops acne, and Maggie's eyebrows grow shaggy and she develops super strength after drinking a cheap milk option from the Kwik-E-Mart that is based on "Science."[17] A narrative drove the premature outrage, but the lasting ramifications were driven by business and politics.

Seeing an opportunity for differentiation in the marketplace, Oakhurst Dairy in Maine became the first dairy operation to incentivize their farmers not to use rbST to increase milk production. The farmers would sign a pledge not to use additional hormones in their operations, and Oakhurst could label their milk "rBST free," laying claim to "America's first Farmers Pledge" against rbST.[18] The competitive advantage wouldn't last long as more companies added disclaimers to their milk labels like, "This milk is from cows not treated with rbST." Some labels went so far as to say "Hormone Free," but this was inaccurate since naturally occurring bST is present in all cow's milk, as we have already stated.

So, what does the data show? Does it support the anti-rbST contingent's position that the synthetic hormone poses a significant risk to human health? We have already stated that the FDA determined the product to be safe for humans, but even Canada's regulatory bodies that banned the use of rbST in the country did not reject the synthetic hormone because of its direct effect on human health. They worried that the hormone could increase the incidence of mastitis in dairy cows, resulting in contamination of the milk from bacterial infection. There is absolutely no support to the notion that rbST poses a risk to humans.

Actually, milk producers are required to include the statement, "The Food and Drug Administration has determined that there is no significant difference in milk from cows treated with rbST and non-rbST cows" if the company chooses to include

food for thought... (cont'd)

language like "rbST-free" on the label.[19]

However, the advocacy against its use and the market demand for rbST-free milk essentially reduced the use of the synthetic hormone in nearly all dairy operations across the county. It is even banned in places like Canada, Australia, and the European Union. Today, it is hard to find milk that would contain additional hormones like rbST, but the label remains on many products because of the peace of mind it provides consumers who were influenced by the marketing campaigns.

And how about the idea that hormones in our milk are causing precocious puberty, the initial reason for this Food for Thought? According to the medical community, rbST is not the culprit. While there is speculation of a causal effect from endocrine disrupting chemicals, like plastic components, and even a correlation to obesity rates, they are still just that: speculation. While some causes, like thyroid disorders, certain types of tumors, and a few genetic conditions have been identified, in 90 percent of the cases in girls the cause remains unknown.[20] However, the premature outrage over a new technology has removed a tool from our toolbox to meet the challenge facing food production today.

What about crop rotations? Advocates who tout the benefits of organic agriculture use an ever-so-subtle sleight of hand when it comes to this subject. These advocates highlight crop rotation as a beneficial practice of organic farming while in the same breath castigate conventional agriculture as a horrible monoculture system, clearly insinuating that crop rotations and monocultures are two sides of the same coin. The only problem is that they are not mutually exclusive; they refer to two completely different practices.

For clarity, a monoculture is when a single crop is planted at a time within a field. The primary alternative to monocropping in field crops is intercropping, which refers to the planting of more than one crop in the same general space. Intercropping is utilized most often in agroforestry systems, or in less modernized agricultural systems around the world, particularly in the tropics. It is not a common practice in either conventional *or* organic farming operations in the United States or other industrialized countries.

On the other hand, crop rotations refer to the practice of planting different crops in sequence from one year to the next. I hate to break it to you, but this is certainly not exclusive to organic farms. According to the USDA Economic Research Service (ERS), somewhere between 82 and 94% of conventional acres are in crop rotation![21] With this new perspective, one can conclude that organic and conventional farming systems are not that different when it comes to monocultures and crop rotations.

So, what about the other benefits attributed to organic practices? Are they exclusive to organic farming? When it comes to soil structure and soil erosion, I think the organic advocates have a tough hill to climb. The bedrock of the organic movement has been the exclusion of synthetic pesticides, and this includes herbicides. Crop rotations and cover crops can certainly help minimize weed pressure in the field, but mechanical weed control, or the physical removal of weeds, is still the primary method used by organic farmers in the form of tillage. Often, multiple passes are made with tillage equipment before planting, and multiple passes are made with a cultivator between the rows after the crop has emerged.

Aside from the amount of fuel it takes to make that many passes across the field, and the subsequent amount of greenhouse gases emitted, the physical act of tillage is far from being beneficial for the soil. You may remember our conversation about humus in an earlier *Food for Thought*. We discussed that tillage exposes the soil's organic carbon to the air, resulting in substantial releases of CO_2 into the atmosphere. Additionally, tillage disrupts the organic matter and beneficial microbiology in the soil, inhibits water infiltration, and breaks down the physical structure of the soil that enables it to hold water and nutrients for the growing crops.

In contrast, this is an area where conventional agriculture continues to improve. In 2017, no-till or conservation tillage practices were used on just over half (51%) of all cropland acres.[22] While it is not a direct comparison, the closest data we have for organic agriculture is from the 2019 Organic Survey that showed 36% of all organic farms implement some conservation tillage practices on at least a portion of the acreage. I guess if we were to summarize the soil debate, we could say that 1) organic leads in cover crops, but conventional agriculture is gaining in adoption, 2) crop rotation is a predominant feature of both systems, and 3) conventional agriculture has the clear advantage when it comes to soil quality because of the prevalence of conservation tillage practices. If this really were an either/or situation, I think we would have to give it to conventional agriculture on the soil debate.

— At Least It Doesn't Use Synthetic Fertilizers —

The next benefit claimed for organic farming is the retention of nutrients in the soil, essentially claiming that conventional agriculture practices are more prone to leaching nutrients into surface and ground water because of the application of synthetic fertilizers. This one always causes me to take a quick pause because of my previous life as a sales agronomist. Oh boy . . . if they only knew the lengths farmers went to make sure every drop of fertilizer is used by the plant!

Without going too deep into the economics of agriculture, I will just say that profit margins in farming are thin. There are years that farmers are simply trying to not lose too much money so they can stay afloat for one more year in hopes that commodity prices will be better next season, or the weather cooperates, and so on. Farmers are eternal optimists. With that being said, I would be so bold as to say that most farmers are very shrewd businessmen and women. They know what they expect to make from an acre of land, and they want to minimize their input costs as much as possible. They have numerous fixed costs on the farm—rent, taxes, insurance, labor, and equipment, to name a few. The variable input costs, like fertilizer and pesticides, are scrutinized heavily to ensure that every dollar spent has the best possible chance of return.

I lead with that description because I have been in the business of selling fertilizer to farmers. They want to put on the minimum amount that will yield what they expect and not a pound more. Even though the argument on the organic side of the aisle talks about massive amounts of fertilizer being applied to cropland that simply runs right off the field and into the nearest waterway, I just have not found this to be the reality. Of course, there are instances when unforeseen rain events occur after fertilizers have been applied to a field and there is movement of the fertilizer off-target, but this is the opposite of what farmers strive for. Every pound of fertilizer that ends up somewhere other than in the crop is a dollar wasted to the farmer. It is also worth noting that organic sources of nutrients, like manure, would be subject to the same fate under these conditions, resulting in nutrient run-off into the nearest waterway.

But it would be disingenuous to claim that there have never been issues with runoff from farms, or that there has been no pollution to our waterways from agriculture. This was, and continues to be, a legitimate concern in the agricultural industry. However, we have seen tremendous improvements on this front. Let's look at the Chesapeake Bay as an example. The bay is fed by a watershed that covers 64,000 square miles, spans six states, and is home to over 18 million people. But like many bodies of water, pollutants like nitrogen and phosphorus from various sources, including agriculture, urban runoff, wastewater, and septic systems can impact the health of the water.

From 1985 to 2019, the nitrogen load on the Chesapeake Bay from agriculture was reduced from 156 to 119 million pounds, a nearly 24% reduction. While agriculture is still the largest source of nitrogen, the measurable decrease can be contrasted with the increased load coming from urban runoff, increasing from 27 to 40 million pounds over the same time, a 48% increase! The story is similar for phosphorus from agriculture, which reduced its impact by 46% while urban runoff increased 44%.[23] And much of this is due to the topics we just covered: cover cropping, conservation tillage, and an overall better understanding of management practices on the farm. While we still, and will always, strive to do better and protect the environment that we all share, I think we should tip our hats to the farmers that are making measurable differences in this area.

What about the specific claim against "synthetic" fertilizers? Are they bad? I would argue no. I am not sure the circular logic required for the OTA to claim that "synthetic pesticides and fertilizers . . . can deplete the soil of valuable nutrients," when fertilizers, by definition, are applied to supply the soil with the valuable nutrients that the crops need. Crops are grown in, and have an intimate relationship with, the soil. The roots of the crops grow down into the soil where water and nutrients in the soil solution are taken up by the fine root hairs. When these nutrients are taken up by the crop, they are used to build the stalks, stems, leaves, and other parts of the growing plant.

Ultimately, a portion of these nutrients make their way to the harvestable portion of the plant and are removed from the soil when the crop is harvested. And the amount that is removed is no small number. Corn, for instance, will remove approximately 0.9 pounds of nitrogen, 0.37 pounds of phosphorus, and 0.27 pounds of potassium for every bushel of production.[24] If we extrapolate this for the average corn yield in the US (172 bushels per acre in 2020), roughly 155 pounds of nitrogen, 64 pounds of phosphorus, and 46 pounds of potassium are removed from every acre as the grain truck leaves the field and heads to the local grain elevator. This loss needs to be replenished to maintain adequate soil fertility levels, and fortunately, our modern agricultural system does just that. Nutrient recommendations have become a science to account for what the yield goals of a field are, how much of each nutrient will be tied up in the soil, and what level of nutrients remain from the previous crop, ensuring that the precise amount of fertilizer is applied to maximize the production on every acre.

Early in this book when we discussed the advancements in agriculture, we highlighted the fascinating story of nitrogen and how scientists harnessed nature to pull nitrogen out of the air so that it could be applied to crops. Other nutrients are also needed, and sulfur may have one of the best stories for how critical nutrients can be sourced in environmentally friendly ways. While there are numerous sources for sulfur, one that I used in my previous life came from industrial plants. But these plants were not manufacturing sulfur—quite the contrary.

Because of the growing consciousness around air pollution, the industrial sector has taken numerous steps to mitigate its impact on the environment. One step was the deployment of "scrubbers" on the smokestacks that rise above manufacturing facilities and coal-fired power plants. The scrubbers contain different reagents that react with the sulfur dioxide gas that was destined for the atmosphere. The result is a sulfur-containing byproduct that can be recycled and used for other purposes. Some scrubbers use a mixture of limestone and water, resulting in a calcium sulfate byproduct known as gypsum. This is used in the production of drywall, cement, and . . . soil amendments for agriculture! Others use an ammonia reagent that produces ammonium sulfate as a byproduct, a perfect combination of two essential elements (nitrogen and sulfur) that I personally used to fertilize the crops I managed.

Two other major nutrients that crops need are phosphorus and potassium. These are primarily sourced from the Earth, mined from nutrient-rich deposits around the world. Phosphorus is derived from rock phosphate, with China, Morocco and Western Sahara, the United States, and Russia topping the charts as the leading producers.[25] The phosphate rock is cleaned, crushed, and treated with sulfuric acid to form phosphoric acid. This can then be combined with compounds like ammonium nitrate to form soluble fertilizer products. The word *acid* was just used a couple of times, which can be scary. But the phosphoric acid that is produced during this process can take another route instead of being processed into fertilizer—it is also used in soda beverages to give them that "tangy" flavor. Maybe a little less scary now?

As for potassium, this element is mined from the Earth as potassium chloride, or potash. It is produced primarily by Canada, Russia, Belarus, and China, which together account for approximately 80% of all potash production.[26] There is little processing needed for potash as it can be applied directly to fields as a source of potassium. So, just to recap, the primary nutrients that are used to fertilize crops are 1) nitrogen, which is fixed from the air that we all breathe; 2) sulfur, which is being captured as a by-product of manufacturing and power plants to reduce air pollution; 3) phosphorus, which is mined from the Earth and is in a state that we can drink before it is combined with an ammonia source and made into the final fertilizer product; and 4) potassium, which is mined directly from the Earth and applied back to the soil the way it came out. Why are these "synthetic" fertilizers bad again?

There are also minor elements, like zinc, copper, manganese, and molybdenum that crops need in smaller quantities. These micronutrients are manufactured into liquid formulations that the plant can absorb through the roots and leaves. One would assume that these would also be "bad" actors given that they are synthesized into specific formulations, but interestingly, many are allowed in organic farming. If you are asking why these synthetic fertilizers are okay when others are not, you are asking the right question.

So, how do organic farmers get the nutrients they need in their fields if they don't use these horrible "synthetic" fertilizers (except the ones that they do use)? According to organic guidelines, the key characteristic is that the nutrients must come from an organic source, meaning from a living animal or plant, instead of an inorganic source, like the potash that is pulled from the ground to be put back on the ground. There are two questions that I think are important regarding this subject. The first is: "Why is this distinction important to adhere to within the philosophical stance of organic farming?" The second is: "Are there benefits of organic fertilizer sources over synthetic, or inorganic, fertilizer sources?"

I am honestly puzzled by the first question. The plant does not care where the atom of nitrogen or potassium comes from, as it is just that: a single atom that is used to build the growing plant. As to the second question, I understand that organic nutrient sources often come in forms that need to be broken down by soil microorganisms. This means they are released more slowly to the plants, can potentially feed soil microorganisms better, and even add organic matter to the soil. Honestly, I would agree that the most ideal nutrient source for plants would be many of these organic sources. But from a *productivity* standpoint, the amount of nutrients contained in most organic fertilizer sources is significantly lower and therefore needs to be applied in much greater quantities. I certainly see benefits, of which I am about to discuss, but there are also considerable downsides.

For the organic farmer, the primary nutrient sources are animal manure, compost, and cover crops. We will not go through cover crops again except to say that they can help assimilate the nutrients in the soil and make them more available for the next crop. (They can also fix nitrogen from the soil as in the case of leguminous crops like certain types of beans or clover.) But I do want to touch on both animal manure and compost.

When it comes to manure, I will gladly admit that this is an ideal nutrient source for crops. Even conventional farmers will use manure whenever they have access to it. It can be a great source of multiple essential nutrients and can also add organic matter back into the soil. But just because something is good doesn't mean that the alternative is bad—or even worse, in this case, that it should be prohibited. It really doesn't even mean that it is realistically feasible to use this better option in every circumstance. A quick exercise might help explain this point better.

An estimated 6.1 billion pounds of nitrogen and 1.8 billion pounds of phosphorus were produced from all sources of manure in 2007.[27] That is a lot of animal poop! But if we look at the acres of corn and soybean planted that year, 86.5 and 63.6 million acres respectively, and then refer to the average corn and soybean yields from that year, 151.1 and 41.2 bushels per acre respectively, a picture starts coming into focus.[28] University data has determined the amount of nitrogen and phosphorus needed to grow a bushel of corn, and the amount of phosphorus needed to grow a bushel of soybean. For 2007, the national corn crop would have required 11.8 billion pounds of nitrogen, or roughly 1.9 times the amount produced from all manure sources. The picture for phosphorus is even worse, as the corn and soybean crop together would have required 6.9 billion pounds, or 3.75 times the amount produced from manure. And these are just two of the crops we grow in the US. While manure is a great source of plant nutrients, and any farmer will utilize it if possible, there is just not enough to go around.

But there should be enough for the organic operations, and I think that is a great thing for those farmers. Particularly if the organic operations can incorporate animals into their production systems to create a closed loop for nutrients on the farm. Whatever is left over, the conventional farmers are certainly there with open arms to take all they can get. This is again one of those instances where either/or thinking can blind us to the reality on the ground. The only way that our total agricultural production could be produced only with organic nutrient sources would be to significantly increase the size of animal agriculture in this country, and I'm pretty sure that is not what the organic advocates want. Rather, the predominant sentiment of this contingent is that we need less animal agriculture because it is inhumane and it is also bad for the environment. Maybe this will change the perception of organic agriculture being superior to conventional agriculture for using manure as a nutrient source. While conventional farmers will use it when available, there is just not enough to go around. So instead, they do the best with the sources they have.

Now that we've covered manure, we're left with compost, the second major source of nutrients for organic farmers. Compost is great, right? Some of us may even have a compost pile in our backyard for our home gardens. Throw in your food waste from fruits and vegetables, toss in some eggshells, some coffee grounds and yard clippings, and you're off to the races. But this often looks quite different in the agricultural setting. In large-scale operations, animal manure is the primary ingredient for agricultural compost. We've already discussed the limitations on the amount of available manure, but set that challenge aside for a moment. Instead, we need to ask: Why is compost viewed as a superior nutrient source for our crops? As with manure, the benefits of compost are initially obvious. Multiple nutrients, additional organic matter, even a reduction in the odors, pathogens, and insects that can be a problem with manure. But what might we not be considering?

A by-product of the composting process, as it turns out, is greenhouse gas (GHG) emissions. Animal manure already contributes to GHG emissions, but the composting process just exacerbates it. If we assume that an average crop will require five tons of compost per acre, research has shown that this would represent a carbon footprint of 10,833 pounds (CO_2 equivalent).[29] Dr. Steve Savage, a food and agricultural consultant, provides a unique take on what this number really means by providing other common examples that would have the same carbon footprint:

- Driving a car for 13,982 miles (at 25 mpg)
- Growing, handling, and transporting 9,641 pounds of bananas from

Costa Rica to Germany

- Producing 985 pounds of beef
- The complete carbon footprint of producing 5.7 acres of conventional corn (including everything from the fertilizer to crop protection chemicals to the seed, fuel, etc.)

While the main thesis of Savage's article was to contrast basic open-air composting (like in your backyard) with anaerobic digestion (which is a highly specialized air-free composting process that uses manures and other waste products as a renewable energy source while providing a by-product that can be used for fertilizer), he makes an important point. Once again, a management practice that is used in organic agriculture is perceived as beneficial by default even though it has negative environmental consequences. We saw the same for soil health, even though organic agriculture utilizes tillage at much higher rates than conventional. For nutrient sources, compost and manure are touted as the only solution even though the only way to have enough organic sources would be to increase animal production. And now composting, which produces an incredible amount of greenhouse gas when used on an industrial scale. Maybe it is not an either/or after all.

— A Focus on a Sustainable Future —

A few closing comments on the "organic is better" discussion before we move to the next section. In my opinion, the allure around organic food and farming is largely driven by the "naturalistic fallacy." This term was coined by the early twentieth-century British philosopher G.E. Moore in his book *Principia Ethica*.[30] Essentially, he states that humans have a fallacy that goodness, from an ethical position, can be defined in naturalistic terms. In other words, because something exists in nature it is good, which can also inversely be translated as anything that does not exist in nature is bad. That sounds a lot like the arguments for why organic food is "better." Conventional farming uses pesticides that were produced in labs, not nature. Never mind the fact we can show that the way we use pesticides is not harmful to humans, that they are strictly regulated to ensure any hazard is excessively below a tolerance level for risk, and that many chemicals are based upon naturally occurring compounds (and are in some instances chemically and molecularly indistinguishable from their natural cousins). The mere fact that humans had a hand in creating them means they are in some way "not good."

If we are to have an honest conversation around the challenges facing humanity regarding food production, population growth, resource limitation, and environmental impact, we first need to check our philosophical approach to the ideas presented in this section. There are some simple truths that need to be a part of framing the challenges. The global population is growing and expected to hit ten billion people by 2050. An increasing portion of the global population is moving from impoverished to middle-class living standards and is requiring better access to natural resources and food availability. There is growing pressure on local, natural resources, ranging from water availability to habitat conservation, to nutrient limitations that we harvest from the Earth. And there are the ever-present concerns about the impact humans are having on the natural environment.

When we accept these truths, one obvious realization is that there is a need to increase food production, but it must be done in a way that is most efficient so that we minimize the impact on the other factors mentioned above. This is why I advocate for the best methods to produce food, wherever they come from, and shy away from ideologies, like those that seem to be more prevalent in promoting organic agriculture. I don't think that organic farming practices are necessarily bad. In fact, I think the limitations imposed on organic farmers have forced them to innovate new and, in some cases, better farming practices that should be considered by conventional farmers. However, we cannot view it as an either/or decision. Rather, we must view it as the most sustainable way to feed a hungry planet.

The yield limitations of organic agriculture that we illustrated by highlighting the additional land it would require to meet current production levels is the crux of the issue—the strict practices required for organic farming are incompatible with our need to feed this planet. Even if it were the best way to produce food, it would simply not be possible on the scale needed for our global agricultural demand. The negative environmental impacts that would result from the increased land dedicated to agricultural production would far outweigh any benefit, even if one did exist. As Stefan Wirsenius from Chalmers University of Technology in Sweden concludes, "There is a huge downside because of the extra land [required] . . . If we use more land for food, we have less land for carbon sequestration. The total greenhouse gas impact from organic farming is higher than conventional farming."[31]

So again, it is not an either/or decision if we are honest with ourselves and reject the naturalistic fallacy that permeates our cultural subconscious. We need to maximize production while minimizing environmental impact. Practices that have been pioneered by organic farmers, like cover crops and more intentional crop rotations, should be evaluated as part of conventional operations. A renewed focus on improving soil health is critical to sustaining our food production systems, focusing on no-till and conservation tillage practices that are already being adopted in many conventional farming systems. But we must also continue to develop new pesticides that are favorable from the perspective of specificity and persistence while minimizing the yield reduction caused by weeds, insects, and diseases.

We don't need the fearmongering that is part and parcel of the marketing efforts associated with the organic movement. We don't need the protests and outrage against manufacturers striving to improve our current production systems and practices. We don't need corporations catering to the "organic activists" and making business decisions out of fear of being boycotted instead of making decisions based on science and data. If you need an example, look no further than the YouTube video called "New MacDonald," sponsored by Only Organic, which exploited children to propagate every negative stereotype imaginable regarding our conventional farming systems.[32]

We can be better. We must be better. And it will take every innovation and technology available to us. This includes the advancements in our understanding of the genetic makeup of the crops we grow and the foods we eat. Crop improvements at the genetic level are quite possibly our best chance at addressing the global food production challenges facing us today, and it happens to be what we will discuss in the next section. See you over there!

1 Smith-Spangler C, Brandeau ML, Hunter GE, Bavinger, JC, Pearson M, Eschbach PJ, Sundaram V, Liu H, Schirmer P, Stave C, Olkin I, and Bravata DM (2012). Are organic foods safer or healthier than conventional alternatives? *Annals of Internal Medicine 157*(5), 348-366. https://doi.org/10.7326/0003-4819-157-5-201209040-00007

2 Hurtado S, Tresserra-Rimbau A, Vallverdú-Queralt A, and Lamuela-Raventós RM (2017). Organic food and the impact on human health. *Critical Reviews in Food Science and Nutrition 59*(4), 704-714. https://doi.org/10.1080/10408398.2017.1394815

3 Eisinger-Watzl M, Wittig F, Heuer T, and Hoffmann I (2015). Customers purchasing organic food – do they live healthier? Results of the German National Nutrition Survey II. *European Journal of Nutrition & Food Safety 5*(1), 59-71. https://doi.org/10.9734/EJNFS/2015/12734

4 Miller HI (2020). *How organic farming exploits consumer demand for 'authenticity.'* Henry I. Miller M.D. https://www.henrymillermd.org/24294/how-organic-farming-exploits-consumer-demand-for

5 FAO (2021). *Organic agriculture FAQ: What are the environmental benefits of organic agriculture?* Food and Agriculture Organization of the United Nations. http://www.fao.org/organicag/oa-faq/oa-faq6/en/

6 Haumann B (2021). *Environmental benefits of organic.* Organic Trade Association. https://ota.com/organic-101/environmental-benefits-organic

7 USDA NRCS (2019). 2020 cover crops insurance and NRCS cover crop termination guidelines. United States Department of Agriculture Risk Management Agency. https://www.rma.usda.gov/en/News-Room/Frequently-Asked-Questions/2020-Cover-Crops-Insurance-and-NRCS-Cover-Crop-Termination-Guidelines

8 Zulauf C and Brown B (2019). Cover Crops, 2017 US Census of Agriculture. *farmdoc daily 9*(135). https://farmdocdaily.illinois.edu/2019/07/cover-crops-2017-us-census-of-agriculture.html

9 USDA NASS (2020). *2019 Organic Survey.* 2017 Census of Agriculture Special Studies 3(4). United States Department of Agriculture National Agricultural Statistics Service. AC-17-SS-4 https://www.nass.usda.gov/Publications/AgCensus/2017/Online_Resources/Organics/ORGANICS.pdf

10 Ibid.

11 Hellerstein D, Vilorio D, and Ribaudo M (Eds.) (2019). Agricultural resources and environmental indicators, 2019. *Economic Information Bulletin 208.* United States Department of Agriculture Economic Research Service. https://www.ers.usda.gov/webdocs/publications/93026/eib-208.pdf?v=8521.3

12 Precocious puberty (2021). Diseases & Conditions. Mayo Foundation for Medical Education and Research (MFMER). https://www.mayoclinic.org/diseases-conditions/precocious-puberty/symptoms-causes/syc-20351811

13 Sampson H (1695). A Relation of One Hannah Taylor, a Very Extraordinary Child of about Six Years of Age, Who in Face, etc. Was as Large as a Full Grown Woman; and of What Appeared on the Dissection of Her Body: By Dr. Hen. Sampson, Fellow of the Colledge of Physicians, London. *Philosophical Transactions (1683-1775) 19*, 80-82. http://www.jstor.org/stable/102283

14 Bonner SD (2012). Looking into our childrens eyes: Precocious puberty-adolescents before their time. *Journal of Youth Ministry 10*(2), 7-18.

15 Herman-Giddens ME, Slora EJ, Wasserman RC, Bourdony CJ, Bhapkar MV, & Koch GG (1997). Secondary sexual characteristics and menses in young girls seen in office practice: a study from the Pediatric research in Office Settings network. *Pediatrics 99*(4), 505-512.

16 Posilac (2021). *Union Agener.* Union Agener Animal Health. https://www.unionagener.com/posilac/

17 Bovine Somatotropin (bST) (2021). *Product safety information.* U.S. Food & Drug Administration. https://www.fda.gov/animal-veterinary/product-safety-information/bovine-somatotropin-bst

18 Welk-Joerger N (2016). *Milk moustache misconceptions*. Nicole Welk-Joerger Blog. https://welkjoerger.com/2016/01/13/milk-mustache-misconceptions/

19 Cunnane CB (2018). *Are there hormones in milk? Dirt to Dinner*. https://www.dirt-to-dinner.com/are-there-hormones-in-milk/

20 Duke Health Blog (2020). When is puberty too early? Duke University Health System. https://www.dukehealth.org/blog/when-puberty-too-early

21 Wallander S (2013). *While crop rotations are common, cover crops remain rare*. Amber Waves. United States Department of Agriculture Economic Research Service. https://www.ers.usda.gov/amber-waves/2013/march/while-crop-rotations-are-common-cover-crops-remain-rare/

22 USDA NASS (2019). *United States Summary and State Data*. 2017 Census of Agriculture Geographic Area Series 1(51). United States Department of Agriculture National Agricultural Statistics Service. AC-17-A-51. https://www.nass.usda.gov/Publications/AgCensus/2017/Full_Report/Volume_1,_Chapter_1_US/usv1.pdf

23 ChesapeakeProgress (2021). 2025 Watershed Implementation Plans (WIPs). Chesapeake Bay Program. https://www.chesapeakeprogress.com/clean-water/watershed-implementation-plans

24 Silva G (2017). *Nutrient removal rates by grain crops*. Michigan State University Extension. https://www.canr.msu.edu/news/nutrient_removal_rates_by_grain_crops

25 Pistilli M (2020). *10 top phosphate countries by production*. Investing News. https://investingnews.com/daily/resource-investing/agriculture-investing/phosphate-investing/top-phosphate-countries-by-production/

26 Natural Resources Canada (2021). *Potash Facts*. https://www.nrcan.gc.ca/our-natural-resources/minerals-mining/minerals-metals-facts/potash-facts/20521

27 US EPA (2020). *Estimated animal agriculture nitrogen and phosphorus from manure*. United States Department of Agriculture Environmental Protection Agency. https://www.epa.gov/nutrient-policy-data/estimated-animal-agriculture-nitrogen-and-phosphorus-manure

28 USDA NASS (2008). *Crop Production: 2007 Summary*. United States Department of Agriculture National Agricultural Statistics Service. Cr Pr 2-1 (08). https://downloads.usda.library.cornell.edu/usda-esmis/files/k3569432s/dr26z061h/jh343v888/CropProdSu-01-11-2008.pdf

29 Savage S (2013). *The shocking carbon footprint of compost*. Applied Mythology. http://appliedmythology.blogspot.com/2013/01/the-shocking-carbon-footprint-of-compost.html

30 Baldwin T (2010). *George Edward Moore*. (Zalta EN, Ed.). The Stanford Encyclopedia of Philosophy: Summer 2010 Edition. https://plato.stanford.edu/entries/moore/

31 Varanasi A (2019). *Is organic food really better for the environment?* State of the Planet, Columbia Climate School. https://news.climate.columbia.edu/2019/10/22/organic-food-better-environment/

32 Only Organic. (2015, Feb 22). New MacDonald [Video]. YouTube. https://www.youtube.com/watch?v=ypF15z3euwM

Part Two

Non-GMO Verified

Brace yourselves—this may be the most polarizing topic that we discuss together. I mean, what other topic actually has a moniker like "frankenfoods"! But what do those three little letters even mean? As scientists have made tremendous advancements in biology and medicine, we have reaped the benefits of gaining a better understanding of how our world works—and all the pieces in it. All of this has led to technological advancements in the field of genetic engineering and, ultimately, to our friends the GMOs, or genetically modified organisms.

But what does that mean, a genetically modified organism? Honestly, it does sound kind of scary. We need to distinguish between terms that are often used interchangeably but inherently carry a different meaning. To start, genetic modification is an extremely broad term that simply means the alteration of the genetic composition of an organism to produce desirable traits. In other words, if the resulting progeny (babies) possess desirable qualities or characteristics due to an outside influence, genetic modification has occurred. Can anyone say labradoodle? That's right, humans have a storied history of genetically modifying our food, our livestock, and even our pets, dating back millennia!

But that is not the definition that most people have an issue with. We need to keep going down the list. How about genetic engineering? According to the definition provided by the USDA, it is the "manipulation of an organism's genes by introducing, eliminating or rearranging specific genes using the methods of modern molecular biology, particularly those techniques referred to as recombinant DNA techniques."[1] Okay, but what the heck are "recombinant DNA techniques"? The USDA defines them as "procedures used to join together DNA segments in a cell-free system . . . [that] can be introduced into a cell and copy itself." In layperson's terms, it is the way genetic material from one cell is transferred into another so that it can become part of the recipient's DNA.

Whew, ok. Now there's just one final definition we need to cover before we dive into the details, and that is *transgenic organisms*. These are organisms resulting "from the insertion of genetic material from another organism using recombinant DNA techniques." That's the one we were looking for! This is where the fun name "frankenfoods" comes from, and it is a topic we are going to discuss at length, from the science to the controversy. But let's not get ahead of ourselves.

I think it is important for all of us to have a basic understanding of the terminology, processes, and the way we use new technologies in our food production systems today. In the introduction of this book, I mentioned that we are on the cusp of a genetic revolution, and this was not meant to be hyperbolic. We are advancing in the field of genetics, genomics, proteomics, metabolomics, transcriptomics, and every other "-omics" you can think of. What this means for us as consumers is that we are going to be the beneficiaries of scientific advancements that will continue to make food more readily available for a growing population, as well as making it more efficient to grow and more nutritious to eat. However, if we allow ourselves to be influenced only by the fear that surrounds new technologies in our food systems, we will ultimately face the repercussions of stopping scientific progress in its tracks.

To make sure we, as consumers, have the knowledge and ultimately the power to decide for ourselves what food we are comfortable with putting into our bodies—and moreover, what criticisms we need to push back on—we need to understand how our food is produced. We must be informed so as not to be influenced. Let's spend some time together in this section and go through the history of plant genetics and breeding, discuss the controversies that surround the GMO debate today, and take a glimpse into what the future has in store for us.

1 USDA (2021). Agricultural biotechnology glossary. United States Department of Agriculture. https://www.usda.gov/topics/biotechnology/biotechnology-glossary

Non-GMO Verified:
A Few-Thousand-Year Origin Story

— Humans Discover Selective Breeding —

I intentionally mentioned labradoodles earlier to call out a specific type of genetic modification that we don't give a second thought to these days: selective breeding. Another name for it is artificial selection, and both terms were coined by noted evolution theorist Charles Darwin. I brought up labradoodles in jest, but dogs are thought to be the earliest example of genetic modification in human history, dating back to our hunter-gatherer ancestors in east Asia. The current theory is that wild wolves began to follow the nomadic people around as scavengers and were eventually domesticated. Our ancestors started to select the wolves that were more docile and amenable as "house pets." Eventually, other characteristics were selected, such as height, body type, hair color and length, basically all the features that would make them better suited for the jobs the people had in mind for their new best friends. While the American Kennel Club currently registers 195 different dog breeds, they can all trace their ancestry back to the wild wolf—and they can thank our early ancestors for their chic new looks.[1]

The first examples of genetic modification in food crops are thought to have happened much later, which would make sense since hunter-gatherers rarely settled down long enough to reap multiple harvests and select crops that were "the cream of the crop." See what I did there? While not the first example of genetic modification, one of the most well-known examples is the creation of corn as we know it today. Creation? Yes, the crop that is currently planted across ninety million acres in the United States[2] and more than 450 million acres around the world[3] would not be here if not for direct human intervention and genetic modification.

Corn traces its ancestral roots back 7,000 years to a plant called teosinte, a wild grass native to Mexico. Looking at the two plants side-by-side today, you would be hard-pressed to see the familial resemblance.[4] What we know as the cob and kernels on today's corn plants were barely more than oversized seed heads emerging from numerous offshoots of the ancestral grass. But the early Mesoamerican people relied on this plant for food and began selecting for larger kernels. Eventually, the structure of the plant began to change. Larger kernels meant a larger seed head, which required a more robust architecture to hold it up. It's hard to believe today, but Christopher Columbus also "discovered" corn when he discovered America (or, more accurately, the Bahamas archipelago and the island of Hispaniola, i.e., modern-day Haiti and the Dominican Republic). This crop was not known to Europeans at the time.

Corn is not the only example of where human intervention aided in domesticating a wild plant and transforming it into a more productive food crop. There are numerous examples of common foods that have changed dramatically from their ancestral relatives, including watermelon, bananas, eggplant, carrots, and peaches to name a few. And in every example, humans have made significant "improvements" to these food crops by selecting for the desirable characteristics that we enjoy. All of this was based on the principle of genetic modification. Again, the definition of genetic modification is applying selection pressure to achieve desirable characteristics, ultimately changing the genetic makeup of the organism. Every one of these examples is a clear illustration of that process at work. Smaller seeds, larger fruit, more edible flesh. All characteristics that humans deemed desirable, and hence, selected the plants that yielded these qualities to use as the seed stock for subsequent plantings.

The last example I like to point to is the diversity of cole crops, a term used to describe members of the Brassica family that we consume today as part of our daily servings of vegetables. Looking at broccoli, kale, brussels sprouts, cabbage, cauliflower, and kohlrabi, it is easy to distinguish between them based on their unique characteristics, but each of these veggies are simply different varieties of the same *species* of plant! They all share the same binomial nomenclature (genus and species), *Brassica oleracea*, and originated from the same ancestor.

That means that from a single wild mustard plant in the Mediterranean, humans began to select for different characteristics—like stems, lateral leaf buds, or flower buds—to create six unique varieties of the same plant. [5] While our hunter-gatherer ancestors, ancient Mesoamericans, or even the early wild mustard consumers in the Mediterranean didn't understand the principles of genetics or inheritance, they were unknowing participants in artificial selection—and they pioneered the way for future applications of modern genetic engineering.

— The Field of Genetics is Born —

You are going to have to bear with me as I discuss the first real breakthrough in our understanding of how genetic inheritance functioned. My first scientific passion was the field of genetics, and the story we are about to uncover is one that is foundational to any geneticist. But before we get into the science, let's paint the backstory.

The breakthrough would come from what may seem like an unlikely source. In the early 1800s, a poor farmer's son, Johann Mendel, would leave his family farm in Austria to pursue an education, ultimately enrolling at the Philosophical Institute of the University of Olmütz. Struggling to survive on the little bit of money he could make tutoring other students, and then suffering from bouts of depression, Mendel returned home. However, he would not accept the fate of taking over the family farm. Instead, he graduated from the University and joined the Augustinian order at the St. Thomas Monastery in Brno, taking on his more recognizable name of Gregor.[6]

Mendel's duties as a priest to console the sick and dying brought back his struggle with depression, so he was moved into a teaching role, only to fail the teaching certification exam. The monastery decided to send him to the University of Vienna to continue his education in the sciences. Mendel thrived in this new environment, working with prestigious scientists studying physics, mathematics, and the anatomy and physiology of plants.

Mendel returned to the monastery and was eventually elected abbot. In addition to his administrative duties, he began an experimental program to investigate hybridization in plants (basically the process of producing offspring by mating different varieties or species) and the transmission of hereditary characteristics. The prevailing theory of hybridization at the time was that the offspring of the initial hybrids, which were formed from crossbreeding, would revert to the original characteristics. Mendel chose edible peas as his test subject and examined thirty-four varieties for seven traits that were distinctive and easy to measure, consisting of what Mendel referred to as "character-pairs." I won't bore you with the entire experimental design or the methods he used, but Mendel essentially crossed pea varieties that differed in the character pairs—such as a tall and short variety, or one variety with green seeds and one with yellow seeds. What he observed would forever change our understanding of genetic inheritance.

Mendel began to observe ratios in the offspring that were repeatable in different generations. He first crossed the parent plants and found that all the offspring in the first generation (the F_1 generation) had the characteristics of one of the parent varieties. Next, he crossed all of these identical plants together, called self-pollination, and found the resulting progeny produced characteristics in a ratio of 3:1, where the other characteristic was observed again (the F_2 population). Based on this observation, Mendel described the characteristics as either dominant or recessive. He continued to self-pollinate the offspring for additional generations and again found ratios, like 1:2:1 in the third generation where 50% were hybrids (having one dominant and one recessive gene) and 50% were "true-breeding," of which half were dominant (having two dominant genes) and half were recessive (having two recessive genes). See the figure below for a visual representation of the ratios that arose with each generation.

Mendel's approach to experimental design is what provided him the opportunity for success. Firstly, he followed individual characteristics separately, which allowed him to statistically analyze the inheritance patterns. Secondly, he used large populations so that the ratios would become evident. Although not formally named until much later, two of the key findings from Mendel's work have been called the law of independent assortment, whereby characteristics are inherited independently of each other, and the law of segregation, whereby each parental line has a unique "germ cell" capable of contributing a characteristic to the progeny, supportive of the cell theory of fertilization. Like many great scientists, Mendel's contributions to the field were not immediately recognized, but today he is regarded as the "father of genetics" and the type of inheritance he discovered carries his namesake, Mendelian inheritance.[7]

Hopefully that didn't put anyone to sleep, but I think it is fascinating that nearly all genetic discoveries today can trace their roots to the simple pea experiments conducted by an abbot in an Austrian monastery in the 1800s. It is also important, in my opinion, to look back at history and understand the origins of any scientific inquiry. While we wrestle with the concept of genetically modified foods today, the scientific discovery that would lead to this innovation started nearly two hundred years ago. And while the current debate seems to have erupted with the introduction of the FLAVR SAVR tomato in 1992, followed by pesticide-tolerant crops a few years later, genetically modified organisms were a hot-button issue two decades earlier when two scientists successfully created the first genetically *engineered* organism.

Illustration of an example pea cross from Mendel. Cross between two distinct parents results in all progenies identical to one parent. Self-crossing this generation and subsequent generations informed Mendel of the hereditable characteristics that would reveal the existence of dominant and recessive traits. A Punnett square is also provided to show how self-crossing F1 progeny would result in a 3:1 ratio based on observable characteristics, and a 1:2:1 ratio based on genotypic characteristics.

— From Modification to Engineering —

As previously mentioned, the term genetic modification is extremely broad and can cover everything from our labradoodle example to Mendel's pea plants to the most current and advanced methods of breeding that we deploy today.

However, most people are concerned about the concept of genetic engineering, and more specifically transgenic organisms. So, let's go back to when the first one was created!

In 1973, Herbert Boyer and Stanley Cohen successfully transferred antibiotic resistance into one bacterium by inserting DNA that had an antibiotic-resistant gene from another bacterium.[8] What a discovery! This checks off all the terminology we just reviewed above. It was certainly *genetic engineering* because genetic material was inserted using *recombinant DNA technology*. And it was the movement of genetic material from one organism to another, so they absolutely created *transgenic organisms*.

This accomplishment was met with all the fanfare that you would expect: immediate blowback and criticism from the media, the government, and even other scientists about what the ramifications of this new technology could be. Wait . . . what? Yes, instead of celebration, there was a moratorium on all genetic engineering work until the science could be sufficiently discussed by the "experts." Actually, it is understandable that a discovery of this magnitude would be viewed through a critical lens. Many questions needed to be answered. Could the transgenic organisms impact human health if not properly contained? Could the genes introduced into bacteria move independently into other organisms? If we genetically altered viruses using this same methodology, would that impact their pathogenicity? One of the fears levied against this breakthrough was the possibility that it would lead to human cloning and what this could mean from a moral or ethical perspective. It may seem like a big leap to think that moving a small bacterial plasmid containing one gene and consisting of only a few thousand nucleotide base pairs into another bacterium would automatically lead to the complete copying and expression of the entire human genome (estimated to have between 20,000 and 25,000 genes and approximately 3 billion base pairs[9]), but that's how significant this discovery was to the field.

Back to the matter at hand. The experts convened in 1975 for the International Conference on Recombinant DNA Molecules, or more commonly known as the Asilomar Conference. During the opening session, David Baltimore (the man who created the classification system for viruses, now called the "Baltimore classification"!), explained to the attendees that they would be expected on the last day to reach a consensus on the guidelines and recommendations for how to move forward with this line of work. Scientists, government officials, and lawyers from around the world debated the issues during a hotly contested three-and-a-half days. At one point, Sydney Brenner of the Molecular Biology Laboratory at Cambridge stated, "Does anyone in the audience believe that this work . . . can be done with absolutely no hazard? This is not a conference to decide what's to be done in America next week; if anyone thinks so, this conference has not served its purpose."[10] By the end of the conference, there was actually great consensus within the scientific community and global governing bodies to continue this type of research. However, researchers would need to capitulate to numerous guidelines, regulations, and restrictions that would be put in place to ensure the public that all precautions were being taken.[11] This would come back to bite them, which we will return to shortly.

Within a few years of Boyer and Cohen's discovery, the first living organism would be patented. We have an entire section coming up called "The Right to Own . . . Life?" so we will hold our bioethical discussion for then. But for now, the important nugget is that a genetic engineer at General Electric created a genetically engineered bacteria that would break down crude oil, providing a useful tool in the event of an oil spill. In a case that went the whole way to the US Supreme Court, GE was awarded the patent in 1980 and paved the way for the ownership of genetically modified organisms that had been created in a lab.[12]

As for Boyer and Cohen, they took two drastically different paths after their discovery. Cohen remained in the university system for the rest of his tenure and continued to pursue basic research in the field of genetics. Boyer, on the other hand, saw the promise of the new recombinant DNA technology and cofounded Genentech just three years after his breakthrough. The first commercial success of this technology for humans would come from a collaboration between Boyer's

Genentech and the pharmaceutical giant, Eli Lilly and Company. Their target: diabetes.[13]

It may be hard to conceive of now, but prior to the 1920s, diabetes was often an early death sentence. It wasn't until then that scientists, specifically Frederick Banting, Charles Best, and James Collip, discovered that insulin could be harvested from the pancreases of cows and injected into diabetes patients.[14] Shortly after, Eli Lilly launched Iletin, the first commercially available insulin product.

Interesting side note, Banting, Best, and Collip patented both the insulin product and the process to make it from beef pancreas and received a whopping $3 for the discovery—between the three of them! According to an article in Clinical Chemistry, "For $1.00 to each, the three discoverers assigned their patent rights to the Board of Governors of the University of Toronto. . . A patent was necessary to restrict the manufacture of insulin to reputable pharmaceutical houses who could guarantee the purity and potency of their products."[15]

Although there was now a process to produce insulin, it was time-consuming and there were fears in the 1970s that the growing number of diabetes patients would outpace the supply of beef pancreases. Bruce Frank, a former scientist for Eli Lilly, recollects, "We were all becoming aware that the curve for the number of diabetics, and therefore the insulin demand from that group, would be crossing over the curve of pancreatic availability somewhere toward the end of the 1990s."[16] Recombinant DNA technology provided a promising alternative, and in 1982, that solution became a reality.

The collaboration between Boyer's Genentech and Eli Lilly proved fruitful as they were able to synthesize human insulin from genetically engineered bacteria. For those interested in the process, the human gene responsible for insulin production was inserted into bacteria that were then placed in large fermentation tanks to reproduce. The large quantities of bacteria produced the insulin, which was then harvested, purified, and made available commercially under the brand name Humulin. This would be the first drug produced by a genetically engineered organism and the first to receive approval for human consumption by the FDA. While there are still controversies over the costs of insulin, and the possibility of price gouging in the market for this drug, it can be said with confidence that the introduction of GMO-produced insulin curtailed any question about the availability of the drug from a purely supply-oriented perspective.

Previously, I mentioned that the conciliation between the scientists and companies involved in genetic engineering research and development at the Asilomar Conference would eventually come back to bite them. Synthetic insulin, it could be argued, is one of those instances. Eli Lilly submitted NDA (New Drug Application) 18-790 for Biosynthetic Human Insulin (BHI) – Humulin to the FDA in 1982. Hank Campbell, writing for the American Council on Science and Health, states that many scientists at the FDA wondered why the drug was being considered a New Molecular Entity because its amino acid sequence is the same as the insulin that occurs naturally in humans. This means that at the most basic level, the drug was not new—rather, it was a chemically indistinguishable version of a naturally occurring hormone. Even so, Campbell claims that the scientists and drug companies remained vigilant because they were "worried about [Ted] Kennedy or others shutting them down," and instead "publicized their efforts to assure the public that recombinant DNA research was not opening up a new world of biological problems." Campbell summarizes that "activists began to promote the belief that scientists were being so cautious because their research was risky . . . That is the legacy they left us. Two generations of activists have convinced a giant swath of Americans—and the politicians that represent them—that science 'must be proven 100 percent safe' before it can be on the market – a test no product can withstand."[17]

This sentiment is clearly biased to one side of the argument, but it does illustrate the classic catch-22 situation that scientists were faced with in the early days of genetic engineering. If you don't go the extra mile to follow all the regulations and restrictions that have been levied on you, we will shut you down. If you do follow them, it only goes to prove that you are working on something that is clearly a threat to society. I can certainly see parallels to today's arguments, but it is worth asking: before reading this section just now, how many of you were aware that the modern insulin used by diabetics is produced by transgenic organisms? Does this change your perspective on the concept of genetic modification and GMOs? It may be something to keep in mind as we continue down this memory lane of genetic engineering.

food for thought...

Time for a gut check, people. We can see that there was significant controversy using recombinant DNA technology, or GMOs, in the field of medicine when scientists first started producing insulin with these techniques. The current controversy surrounding GMOs in food exhibits striking similarities. It's not based on science, per se, but rather a gut feeling—or herd mentality—made more impactful with the connectedness of people through social media.

Opponents can easily generate outrage in a public that doesn't have a deep background in the subject matter by stating some salacious claims, creating a catchy slogan, and then picking a target for the outrage.

But we saw the outrage die off from the use of GMOs in medicine after insulin. I don't hear about it at all, in fact. Is this because scientists saw the error in their ways and stopped using this technology? Absolutely not! To the contrary, genetic engineering is critical in medicine and responsible for saving lives every day.

According to an article in the New Republic, "In 2014, 10 of the top 25 best-selling drugs were . . . made up of recombinantly produced proteins—including blockbuster treatments for arthritis, cancer and diabetes."[18] Regarding vaccines, they state, "Of the 10 vaccines that the Centers for Disease Control and Prevention (CDC) recommends for newborns, three are available in recombinant form." Speaking of vaccines, we have just witnessed perhaps one of the greatest medical feats in history in response to the COVID-19 pandemic to bring multiple effective vaccines to market in record time. Do they fit into this conversation on GMOs? Let's hold that question for later. 🖤

— Coming to a Supermarket Near You —

It may seem like the transition from producing drugs from genetically engineered bacteria to developing genetically engineered plants would be a natural progression, but boy would you be wrong! I kid, but scientists needed to overcome numerous challenges before the breakthroughs we will cover in the rest of this section were even possible. Perhaps the most important came from studies being conducted on the common bacteria, *Agrobacterium tumefaciens*.

The bacteria is responsible for causing crown gall, a tree disease you have probably seen hundreds of times and never paid it a second thought. The scientific name, "tumefaciens" may have given it away, but essentially these bacteria cause cancerous growths, or tumors, in trees. While Agrobacterium had already been identified as the causal agent of crown gall disease, the work of Armin Braun, a plant pathologist at the Rockefeller Institute in Princeton, provided the key that plant breeders would need to make the next advancements in the field of genetic engineering.

Researchers working on crown gall at the time were coming across a perplexing problem—they couldn't always recover the bacteria from the galls, or tumors, on the plants. This was odd because conventional wisdom would presume that the causal bacteria of a lesion, or growth, should be present in the infection site. Additionally, the researchers noted that secondary tumors would form on a plant far away from the primary infection site. Braun verified that while the bacterium was needed to cause the initial infection, it was not needed for the development of the primary or secondary tumors. He even showed that he could graft the tumors that contained no Agrobacterium onto healthy plants and the tumors would continue to grow. (Grafting is a process commonly used in horticulture to attach the stem of one plant, often a tree, onto another so it grows as a single plant. Perhaps one of the most interesting examples of this is the "Tree of Forty Fruit," a creation by the artist Sam Van Aken who grafted forty different types of fruit trees, including peaches, plums, cherries, almonds, and others onto a single root stock![19])

There were numerous revelations about the Agrobacterium infection that Braun discovered, but the punchline for our purposes came when he proposed that there existed a "Tumor Inducing Principle" (TIP) in the bacteria. Essentially, he proved that the bacteria were able to transfer the mechanism for tumor development even though the bacterium itself was not present after the initial infection took place. He proposed four possibilities for how this could occur, including the correct one: "A chemical fraction of the bacterial cell, such as a DNA molecule . . . is capable of initiating in a host cell a permanent developmental alteration."[20]

We can all take our hats off to Braun for his contribution to the field of plant genetics, but it would take another couple of decades before the mechanisms of the tumor-inducing process would be fully understood and could be used for the purpose of plant breeding. Scientists eventually discovered that the "chemical fraction of the bacterial cell" that Braun proposed was, in fact, a secondary chromosome in the bacterium. Agrobacterium still has a set of chromosomal DNA—just like humans, who have twenty-three pairs of chromosomes that provide the blueprint for our every characteristic—but this bacterium also has a second set, called a plasmid. You may be wondering, "So what?" Well, through a complex process, the Agrobacterium attaches to a host cell, injects its plasmid into the cell, and from there the DNA of the plasmid incorporates itself into the DNA of the host plant. The genes carried by the plasmid, including those required to replicate itself, trigger the host cell to provide it with nutrients and ultimately cause the development of a tumor, are now a part of the host's DNA and are expressed by the plant!

I don't know if you are jumping ahead of me, but you may be able to see how this realization sparked the imagination of plant breeders everywhere. Scientists saw this new mechanism, which already existed in nature, as a tool that they could exploit to insert desirable genes into a plant. This could cut down the time involved in selectively breeding plants that had specific characteristics and the accuracy with which they could select for only minor changes. With the introduction of recombinant DNA technology and a mechanism to insert specific genes into the plant, scientists, geneticists, and plant breeders had all the tools they needed to embark on the journey of genetic engineering in the foods we eat.

food for thought...

When *Agrobacterium tumefaciens*-mediated transformations started to make waves, most plant breeding efforts were still based on traditional crossbreeding that focused on selecting the best offspring. However, there was another technique employed that is still being used today in certain food and ornamental crops: mutagenesis. As scientists learned more about the role genes played in expressing certain traits, or characteristics, they were also learning how mutations within a cell's DNA could cause changes in these traits. While most mutations that happen in nature are deleterious and will kill the cell, or even the offspring, some mutations simply alter the way a gene is expressed. This is the basis for natural selection and all the variety we see in the world today.

The problem is that these naturally occurring mutations are very rare and plant breeders may want a trait that is not currently found in nature. As researchers began to understand this process, they realized that they could use chemicals or radiation to cause mutations in the DNA and then screen for plants that now had more desirable traits. This is a gross generalization, but it gets to the point.

The reason you may be interested in this is that radiation-induced mutagenesis is considered a conventional breeding method because no external DNA is being used and, technically, mutations do arise in nature. This means that it is approved under the National Organic Program guidelines. While there are over 3,000 plant varieties in the database maintained by the United Nations Food and Agriculture Organisation and the International Atomic Energy Agency that have been produced using this method (not all are food crops), not a single one would be prevented from displaying the USDA Organic Seal if no chemicals were used in their production. While I am in favor of utilizing all the tools available to us to continue to improve our...

food for thought... (cont'd)

food production systems, it is interesting that there can be outrage over the use of new, precise technologies that introduce desirable genes, or even just alter a single nucleotide in a plant's DNA to have a positive outcome, while we all just go along and eat our delicious Ruby Red grapefruits that have been blasted with ionizing radiation to keep that attractive red color. 🖤

Since this book is supposed to be about food production and agriculture, we should probably transition back to GMOs in our food crops. Genetic engineering first made it into the fields in the late 1980s with the testing of genetically engineered food crops. The first to become commercially available was the infamous FLAVR SAVR tomato that hit store shelves in 1994. This fascinating case study illustrates the need for a convergence between business and science, and how public perception and sentiment can so easily be swayed. So, what was unique about this tomato?

To this day, tomatoes face a considerable problem: getting a tomato from the field to your kitchen without the fruit overripening or softening. The tried-and-true practice involves picking green tomatoes before they can ripen on the vine. This essentially stops the ripening process, which food companies subsequently stimulate again, artificially, through the application of ethylene gas, a naturally occurring, fruit-ripening hormone found in plants. The ethylene restarts the ripening process and turns the tomatoes from green to red, but the process is commonly criticized for not producing the full flavor profile of vine-ripened tomatoes.

It was previously hypothesized that an enzyme in tomatoes, polygalacturonase (PG), which dissolves plant cell walls, was responsible for the softening of the fruits. Davis, CA-based Calgene, Inc. embarked on a project to stop the production of this enzyme, funded initially by a contract with Campbell Soup Co. to keep their tomato paste thicker, another problem caused by the

PG enzyme.[21] However, Calgene turned its attention to the fresh fruit market when Campbell's eventually backed out of the deal, worried that the modification would change its nearly one-hundred-year-old recipe. Through genetic engineering, the researchers were successful in finding a mechanism to prevent the PG gene from being expressed by inserting an antisense, or reversed, copy of the PG gene to essentially cancel it out. After years of field trials, Calgene was able to significantly reduce PG production in tomatoes.[22] From a scientific perspective . . . success!

The initial demand for the FLAVR SAVR tomatoes was high, especially in the targeted launch markets of Davis, CA, and Chicago, IL. Across the Atlantic Ocean, Zeneca licensed the GMO tomatoes to introduce a new tomato paste in the United Kingdom. Again, the initial demand was high and outpaced the sale of traditional tomato paste in a number of stores. However, the genetically engineered tomato would meet an early demise in both markets, albeit for different reasons.

In the United States, Calgene assumed that the reduction in PG would not only prevent overripening and help the tomatoes last longer, but it would also prevent the fruit from softening and bruising during transport. The tomatoes were able to be picked ripe and last in stores, but the second assumption about standing up to shipping did not materialize. Ultimately, the increased cost of production—nearly $10 per pound when the sale price was only $1.99 per pound—became unsustainable and the FLAVR SAVR tomato was promptly retired.

The reason for Calgene's failure is summed up by former Monsanto chief tomato scientist, Harry Klee, who stated, "FLAVR SAVR failed because it made a minimal impact on shelf life/fruit softening, and . . . Calgene chose an old-at-the-time variety that they could access with FTO [freedom to operate]. It was a terrible variety with relatively poor yields. It simply cost them more to make the tomatoes than they could sell them for."[23]

The failure in the United Kingdom was for a much different and, I would argue, much more sinister reason. The tomato pastes from genetically engineered tomatoes saw an immediate reduction in sales after the broadcast of an interview with Dr. Arpad Pusztai and the media firestorm that ensued. Pusztai discussed the results of a study he conducted where he found feeding rats genetically modified potatoes led to adverse biological effects. His conclusion was that the results could be attributed to the overall process of genetic engineering, when in fact the data suggested it was the specific transgene that was introduced. An independent analysis of Pusztai's data concluded that the statements the researcher made in his broadcast were incorrect, but the damage was already done.[24] Zeneca was forced to remove its products from store shelves in the United Kingdom. This example is important to share so that we make sure we don't fall victim to the same sort of baseless attacks on new scientific discoveries.

1 AKC (2021). *Breeds by year recognized.* The American Kennel Club. https://www.akc.org/press-center/articles-resources/facts-and-stats/breeds-year-recognized/

2 USDA NASS (2020). *Acreage (Released June 30, 2020).* United States Department of Agriculture National Agricultural Statistics Service. https://www.nass.usda.gov/Publications/Todays_Reports/reports/acrg0620.pdf

3 Widmar D (2018). *Global acreage: Is the expansion over?* Agricultural Economic Insights. https://aei.ag/2018/04/30/global-acreage-is-the-expansion-over/

4 UCMP (2007). *The other green (r)evolution.* Understanding Evolution. University of California Museum of Paleontology. https://evolution.berkeley.edu/evolibrary/news/070201_corn

5 Maggioni L, von Bothmer R, Poulsen G, and Lipman E (2018). Domestication, diversity and use of Brassica oleracea L., based on ancient Greek and Latin texts. *Genetic Resources and Crop Evolution 65*, 137-159. https://doi.org/10.1007/s10722-017-0516-2

6 Biography.com Editors (2021). *Gregor Mendel biography.* The Biography.com website. https://www.biography.com/scientist/gregor-mendel

7 Olby R (2021). *Gregor Mendel: Theoretical interpretation.* Encyclopedia Britannica. https://www.britannica.com/biography/Gregor-Mendel/Theoretical-interpretation

8 Science History Institute (2017) *Herbert W. Boyer and Stanley N. Cohen.* Historical Biographies. https://www.sciencehistory.org/historical-profile/herbert-w-boyer-and-stanley-n-cohen

9 Pray L (2008). Eukaryotic genome complexity. *Nature Education 1*(1), 96. https://www.nature.com/scitable/topicpage/eukaryotic-genome-complexity-437/

10 Institute of Medicine Committee to Study Decision Making (1991). *Asilomar and Recombinant DNA: The End of the Beginning.* (Hanna KE, Ed.). National Academies Press. https://doi.org/10.17226/1793

11 Rangel G (2015). *From corgis to corn: A brief look at the long history of GMO technology.* Science in the News, Harvard University. https://sitn.hms.harvard.edu/flash/2015/from-corgis-to-corn-a-brief-look-at-the-long-history-of-gmo-technology/

12 *Diamond v. Chakrabarty*, 447 U.S. 303 (1980). https://supreme.justia.com/cases/federal/us/447/303/

13 Science History Institute, Herbert W. Boyer and Stanley N. Cohen

14 Hoskins M (2019). *Way back when…insulin was cheap (and then it wasn't).* Healthline. https://www.healthline.com/diabetesmine/history-of-insulin-costs

15 Rosenfeld L (2002). Insulin: Discovery and controversy. *Clinical Chemistry 48*(12), 2270-2288. https://doi.org/10.1093/clinchem/48.12.2270

16 Philippidis A (2016). *The 'right protein' to establish recombinant therapeutics.* Genetic Engineering & Biotechnology News. https://www.genengnews.com/insights/the-right-protein-to-establish-recombinant-therapeutics/

17 Campbell H (2017). *40 years ago, GMO insulin was controversial also.* American Council on Science and Health. https://www.acsh.org/news/2017/08/29/40-years-ago-gmo-insulin-was-controversial-also-11757

18 Bessen J (2016). GMOs could save your life - they might have already. New Republic. https://newrepublic.com/article/135617/gmos-save-lifethey-might-already

19 Rieland R (2015). A tree that grows 40 different types of fruit. Smithsonian Magazine. https://www.smithsonianmag.com/innovation/a-tree-grows-40-different-types-of-fruit-180953868/

20 Chrispeels MJ and Meins, Jr. F (2014). *Biographical memoirs: Armin C. Braun 1911-1986*. National Academy of Sciences. http://nasonline.org/publications/biographical-memoirs/memoir-pdfs/braun-armin.pdf

21 Philippidis A (2016). *Mistakes shorten first approved GMO's shelf life*. Genetic Engineering & Biotechnology News. https://www.genengnews.com/insights/mistakes-shorten-first-approved-gmos-shelf-life/

22 Bruening G and Lyons JM (2000). The case of the FLAVR SAVR tomato. *California Agriculture 54*(4), 6-7. http://calag.ucanr.edu/Archive/?article=ca.v054n04p6

23 Philippidis A, Mistakes shorten first approved GMO's shelf life.

24 Bruening G and Lyons JM, The case of the FLAVR SAVR tomato.

Non-GMO Verified:
Solutions for Agriculture

— Beginning with Herbicide Tolerance —

While the FLAVR SAVR tomato would not survive in the market, the next round of genetically modified plants would fundamentally transform agriculture. Scientists would take the new tools available and focus on increasing productivity and efficiency in farming. We have already covered the use of pesticides in crop production, but now we will turn our attention to the innovations in genetic engineering that have moved the needle on managing pests even further. The advent of the pesticide-tolerant crop was upon us.

Agricultural statistics are fascinating to study, especially over time. Changes in production, the ability to identify the introductions of key innovations, even the impact of severe weather events on commodity prices and management practices—all of it can have outsized influence in the industry. One statistic that is particularly interesting is the number of farms in the US and the average farm size. Beginning in the mid-1800s, the number of farms began to increase dramatically in the United States. This happens to coincide with the end of the Industrial Revolution and the passage of the Homestead Act of 1862.[1]

The Homestead Act, signed into law by Abraham Lincoln, provided 160 acres in the western territories to any man or woman, black or white, who had not fought against the United States and would agree to settle and farm the land for a minimum of five years and make improvements to the land. The act was consistent with what Lincoln believed to be the purpose of the government: "To elevate the condition of men, to lift artificial burdens from the shoulders and to give everyone an unfettered start and a fair chance in the race of life."[2] We could get into the problems of the law and the unintended consequences that arose, but that is for another discussion. For the purposes of this section, it is merely an explanation for the rapid increase in the number of farms in the US prior to 1900. In total, the Homestead Act, which was only repealed in 1976, resulted in 270 million acres being claimed and settled across much of the western frontier. (Today you can visit the Homestead National Monument in the small Nebraska town of Beatrice. This was where the first homesteader, Daniel Freeman, filed a claim on January 1, 1863.)

The number of farming operations plateaued between the turn of the century and the 1940s, but after World War II there was a precipitous decline in the number of farms that continued for the next several decades. It wasn't until the 1970s that the number of farms stabilized again to the steady, moderate decrease we see to this day. Since the amount of US farmland has remained relatively stable since World War II, farm size has increased at roughly the same rate as the number of farms has decreased, resulting in a continued consolidation in the number of farms.[3]

Farm consolidation is apparent when evaluating the Census of Agriculture, which the USDA conducts every five years. They review numerous trends in US agriculture, one of which is the breakdown of farms by overall size (in acreage). There are seven categories for farm size, but only two of the categories have continued to increase, the smallest farms (1 to 9 acres) and the largest farms (2,000 acres or more). This is because the economies of scale play a significant role in agriculture—most of the crops are commodities and operations are continuing to consolidate to spread the massive costs associated with farming over more acres. The number of farms larger than 2,000 acres grew by 14% between 1997 and 2017, whereas the total number of farms in the US saw a decline of 8% over the same period.[4]

Why is all this important to our conversation? Because farmers have continued to increase the number of acres they farm and their need for greater efficiency in management practices has become ever more important. This is only exacerbated by the seasonality of the agricultural industry. Take soybeans, for example.

Farmers can plant all their acres in an approximately two-week window.[5] These fourteen days may not happen consecutively, as farmers are always subject to Mother Nature's changing dynamics, but even so, planting happens, and must happen, in a relatively short span of time. As farmers push for greater productivity, they must ensure the seeds are in the ground early enough to maximize their yield potential, but not too early to be at risk for a late frost that could kill the crop. It is quite a balancing act!

Because so much of the crop gets put in the ground so quickly, the management associated with raising the crop is also condensed. Early weed control measures, whether pesticides or tillage, need to be undertaken as close to planting as possible. Continued weed control takes place after the crop is growing to reduce competition between the weeds and the crops, but again, the window to accomplish this is relatively small. We can continue to look at other management practices—including fertilizers, insecticides, fungicides, and so on—but the story is the same: the farmer must manage a lot of ground in a relatively short period of time.

In the early- to mid-1990s, farmers were using numerous herbicides to control the weeds that were in their crops. Some herbicides controlled grass weeds, while others controlled broadleaf weeds. Some were put on before the crop was planted and some were applied to a growing crop. There were numerous options available, but the complexity of weed control was certainly high, especially for farmers that were trying to increase efficiency. It's no wonder the new innovations in plant breeding (made possible by genetic engineering) were met with genuine excitement.

Monsanto would simplify the entire process when they released the first herbicide-tolerant crop in 1996. They created soybean varieties that were resistant to the broad-spectrum herbicide glyphosate, more commonly known as Roundup. We covered the controversies surrounding this pesticide earlier, but the reason it is the most widely used herbicide in the world is because of the development and introduction of crops that were tolerant to its application.

So, what made Roundup an ideal target for the genetic engineering crowd? The active ingredient in glyphosate targets an enzyme that is essential for virtually all plants, meaning it has the potential to kill nearly any plant that it is applied to. If researchers could find a variant of the enzyme that was insensitive to glyphosate, then it might be possible to introduce it into crops and make the crops insensitive as well. They would eventually find that enzyme in a bacterium, *Agrobacterium* sp. strain CP4, which was isolated from a glyphosate production facility.[6] The gene for the insensitive enzyme was inserted into soybean lines and did, in fact, confer glyphosate resistance to the soybean plants. Two years after the first glyphosate-tolerant (GT) soybeans were introduced, GT corn would follow and start the broad-scale adoption of glyphosate across much of global agriculture. In addition to glyphosate, crops tolerant to additional herbicides were introduced, and today, approximately 94% of soybean, 91% of cotton, and 89% of corn acres produced in the US are genetically engineered to be herbicide-tolerant.[7]

The broad adoption of GT crops has certainly had positive and negative impacts on agriculture. There have been claims on both sides of the argument, either praising the technology or condemning it, both sides likely go too far in either direction. The one thing that is uncontroversial is that the introduction of GT crops changed farming dramatically. Glyphosate became the most widely used herbicide in the world. As a farmer, it became relatively easy to make weed control decisions because you only needed to make multiple applications of glyphosate throughout the season to control weeds. Even though glyphosate provided virtually no residual control to prevent weeds that would emerge later in the season, the farmer could just continue to make more applications of the product to ensure the weeds were controlled. For the farmer focused on increasing efficiency over a growing number of managed acres, this was a pretty good deal. It also forced other chemical companies to reduce the cost of their herbicides to remain competitive on the acre. However, the "easy button" approach to weed management had ramifications that we are still dealing with today. Without going into a full diatribe, I will simply say that agronomists let down their guard, but we will discuss this in more detail soon. For now, let's simply say that the stage had been set for the proliferation of genetically modified crops in modern agriculture.

food for thought...

The first applications using genetic engineering to manage pests were in weed management. For those not directly involved in farming, I would like to relate the importance of managing weeds to something most of us are familiar with: our lawns. The lawn of the homeowner is a cherished thing. While we get no productive value from it—aside from the intangibles like a place for our children to play, the aesthetics of green contrasting the concrete and blacktop of cookie-cutter neighborhoods, or the increased home value when we go to sell—the homeowner still takes pride in a lush, green lawn free of unwanted grass and broadleaf weeds.

It is estimated that there are roughly 40.5 million acres of lawn in the US.[8] This is equal to approximately 10% of the 390 million farm acres in production every year. While I know we are adamant about the overuse of pesticides in our food production, when broken down by the pounds of pesticides used by different sectors, agriculture used 91% of all herbicides in 2012, while the home and garden sector used 5%. The remainder was used by the Industrial/Commercial/Government sector. Breaking this down to a per acre usage, agricultural land used 1.4 pounds of herbicide per acre and the homeowner used 0.7 pounds, or half of that amount.

Now, keep in mind, herbicides are used in agriculture to make sure that we get the maximum yields to feed our planet. Even with the implementation of other "best management practices," corn production, for example, can result in as much as a 60% yield loss without the use of herbicides.[9] On the other hand, herbicides are used in lawns to make them look pretty. We are about to discuss insecticides in our GMO crops, so how does that ratio stack up comparing agriculture to home and garden use? The 10% of acreage under the supervision of homeowners is responsible for 23% of insecticide use by weight, while the 90% of the acreage in agriculture only accounts for 57%! Maybe we should take a step back and think about the perception of the overuse of pesticides in agriculture today. .●

— Bug Problem? Bring on the Bt Crops! —

The year 1996 was big for GMOs due to the release of glyphosate-tolerant soybeans, but it was also the year that the first insect-resistant crops were released. Where herbicide tolerance was achieved by introducing an enzyme that was insensitive to an application of an herbicide, insect resistance was achieved by enabling the plants to produce their own insecticides. Now that is pretty cool! The insecticide was derived from the bacteria *Bacillus thuringiensis*, more commonly referred to as *Bt*, and the first *Bt*-crops were potatoes, corn, and cotton.

So how were bacteria used to confer insect resistance into plants? We first need to understand how *B. thuringiensis* kills insects. The bacteria produce proteins that are classified as "crystal proteins." When ingested, the high pH of the insect's gut breaks down the protein crystals into a toxin. The toxin then binds to the insect's gut and creates pores, releasing the contents of the gut into the blood and thereby killing the insect.

These toxins are referred to as Cry toxins (in reference to the *cry*stal proteins), and likewise, the genes responsible for encoding these proteins are referred to as *Cry* genes. Perhaps the most important feature of the Cry toxins is their selectivity. First, the process is specific to invertebrates and does not happen in vertebrates, like us. The toxin binds specifically to receptors on the insect's gut cells, and these receptors do not exist in the human gut; instead, the toxin is digested like countless other proteins. But the specificity is even greater. There are countless species of insects in nature. Some of them are pests, but a large majority are actually referred to as *beneficials*. They may be beneficial because they are critical for pollination, because they are predators of other pest insects, or simply because they are not detrimental and should not be eliminated for the risk of disturbing the food chain. As scientists began to learn more about the various Cry toxins, they discovered that each has a limited number of insects, or rather type of insect, for which it is toxic.

Prior to the release of Bt crops, a handful of insect species were causing substantial economic damage and farmers were in a constant battle to protect their crops against these tiny marauders. One such pest was the European corn borer (ECB), or *Ostrinia nubilalis*, which caused tremendous yield loss in corn crops, particularly through the central Corn Belt. The ECB belongs to the insect order *Lepidoptera*, the grouping of insects that contains moths and butterflies. While the adult ECB moths are harmless to crops, the larvae (caterpillar) will chew through leaves and "bore" into the corn stalks, owing to the name.[10] To make matters worse, there can be two, or even three, generations of larvae in a single corn field throughout the season, meaning that damage can occur for prolonged periods, resulting in significant economic losses

This pest was so detrimental that it received the title of "billion-dollar bug," owing to the estimated cost to farmers that resulted from the combination of expenditures on insecticides and the yield loss the insects caused.[11] Numerous studies have quantified this loss, including one from 1995 (the year before the Bt crops were introduced) that reviewed the damage from only one of the affected states. The outbreak that year was so bad in Minnesota that yield losses were estimated to be more than $285 million.[12] Even with insecticides, farmers struggled to control this pest, particularly because of the multi-generational life cycle described above. Additionally, it was difficult to time insecticide applications to contact the insects, which quickly moved inside the corn stalk. When all these factors are taken into consideration, it is clear why the ECB was a prime candidate to target with the use of genetic engineering.

After screening numerous Cry toxins, researchers at Monsanto identified a candidate that provided effective control of ECB: the Cry1Ab toxin. The Cry gene for this toxin was isolated and inserted into the genome of the corn plant using Agrobacterium. For clarity, Monsanto branded its seed varieties based on the toxin they contained, or their trait package. In the case of the ECB-resistant corn that was launched in 1996, they called the trait package Yieldgard Corn Borer. The number of trait packages offered by seed companies today has grown considerably, and often the Bt trait is combined with an herbicide-tolerant trait as well.

Perhaps more important is the increased number of pests that are controlled from the introduction of new Bt proteins. Monsanto's Yieldgard Corn Borer trait package only provided protection from "borer"-type caterpillars, which included the ECB that we have been talking about as well as the sugarcane borer and the southwestern corn borer. However, the next big innovation came with the rapid succession of Cry3Bb1 (Yieldgard Rootworm from Monsanto), Cry34/35Ab1 (AcreMax CRW from Pioneer), and mCry3A (Agrisure RW from Syngenta), launched in 2003, 2005, and 2006, respectively.[13] Each brought protection against the corn rootworm, specifically the western corn rootworm (*Diabrotica virgifera virgifera*), a beetle that by the mid-2000s had become the next "billion-dollar bug." As you may have guessed by the name, the larvae of this insect feeds on the roots of the corn plant, but the adults can also cause damage to the leaves and corn silks.

The adoption of Bt crops grew rapidly, and in the US today, roughly 82% of corn acres and 88% of cotton acres are planted with a Bt crop.[14] One particularly staggering statistic is that since Bt crops were first launched in the late 1990s, over a billion acres have been planted with Bt crops globally![15]

The search for new Bt toxins continues to find new, and often multiple, toxins that are effective against insect pests. Interestingly, a number of Cry toxins have been identified that don't target any known invertebrates, but rather have specific cytocidal ("cell killing") activity against human cancer cells. In these instances, a new nomenclature was used to classify these as parasporins.[16] You gotta love science!

While the first Bt crop was introduced in 1996, the Bt story starts much earlier—all the way back in 1901, on the other side of the world. Insects were dying from an unknown disease, but in this case, it was a bad thing. Large populations of silkworms in Japan were being decimated by a disease referred to as sotto disease, or sudden-collapse disease. It was a Japanese biologist by the name of Shigetane Ishiwatari who first identified and isolated the bacteria that was causing this disease, *Bacillus thuringiensis*. Well, actually, Ishiwatari called it *Bacillus sotto*, but that name would not last.[17]

Ten years later, German scientist Ernst Berliner rediscovered the bacteria in a dead Mediterranean flour moth. Berliner named it after the German town in which the moth was found, Thuringia, cementing the name and identifying a crystal protein in the bacteria along the way.[18] However, the specific activity of the bacteria and crystal protein would not be understood for another four decades.

Even as scientists were working to fully understand the properties of the bacteria, farmers began using it to control insects in their crops. The bacteria were grown and formulated into a dry product that could be "dusted" onto plants. The first recorded use was in France in 1920 and the first commercial product was available eighteen years later under the brand name Sporine.[19] The bacteria proved to be difficult to formulate. Additionally, it would wash away easily with rain, it was degraded quickly by sunlight, it would only kill insects that fed on the surface of the leaves that came into contact with the pesticide, and it was limited to controlling only the larvae of lepidopteran insect species (moths and butterflies). Still, even with these limitations, Bt and its formulations were an important component of insect management for agricultural production.

The bacteria were not used for insect control in the US until the late 1950s, and in 1961, it was registered as an insecticide by the EPA. As time progressed, new strains of the bacteria were identified that controlled dipteran (flies) and coleopteran (beetles) species. However, adoption remained low, partly because of poor formulations, but also because of the introduction of other broad-spectrum insecticides (like DDT) that were significantly more effective against a wide range of insect pests. Once the political and commercial fallout from the anti-DDT advocacy started, however, researchers were driven to find GMO alternatives for pest management, which served as a catalyst for the introduction of Bt crop varieties.

Now that we've covered Bt at length, I have a question for you. Since Bt has been used as an insecticide for decades in the US and abroad, can still be found on the shelves of nearly any home and garden store, and is an approved insecticide for use in organic farming today, why are there objections to using GMO crops that incorporate the bacterial gene into the plant rather than applying the bacteria onto the leaf surface? Is it merely an appeal to the naturalistic fallacy that we discussed earlier?

One of the purported goals of the organic movement was to reduce the amount of pesticides being applied to crops, and GMO Bt crops have certainly done that. In 1996, the year that Bt crops were first launched, there were 19.3 million pounds of insecticides used on US farms. By 2008, the amount had plummeted to 4.01 million pounds, nearly a five-fold decrease![20] There must be something more.

I have scoured the internet and found numerous articles and publications decrying the inherent risks of GMO crops, including those that call out Bt by name. Some blame Bt crops for causing insecticide resistance in insect populations, but this is similar to a discussion we will have regarding Roundup-resistance in a later section, so we will put a pin in that one for now. Some claim that Bt crops are actually dangerous to humans, but I believe (and hope) that this is a fringe argument. We mentioned earlier, albeit briefly, that the Bt toxin is very specific and binds to receptors on the cells of an insect's gut. Humans do not have these receptors, so the toxin is degraded just like any other protein we ingest. There is no scientific merit of fantastical stories of Bt crops causing harm to humans, and even killing them in some cases, so we will not give attention to that claim here. The most prevalent argument against Bt crops appears to be that they are harmful to beneficial and secondary insects, so let's unpack where this objection came from and see if there is a valid concern.

It is simply uncontroversial that the Bt toxin produced by GMO corn plants (and other crop species) is toxic to insects. I mean, the whole point of the GMO crop is to provide protection to the plant by producing its own insecticidal compounds! And that is why Bt is still used in foliar applications in organic agriculture operations because it kills insects that would otherwise harm the crops. The point of contention arises when the risk of harm to non-target insects from an insecticide is presumed to outweigh the benefit it provides from controlling detrimental insect pests. That is the point that became a rallying cry for environmentalist groups at the turn of the millennium.

An example can be found in a 2012 Organic Consumers Association article titled, "Stop GMO Bt Corn Before it Kills Again!" Clearly, this was an objective piece! (Sorry for the sarcasm, I couldn't help it.) It bases its claims on a 1999 research publication that sparked much controversy within the agricultural industry and in the public forum, so it does necessitate some discussion. Let's go straight to the source and review the infamous Losey et al. paper that was published in Nature in 1999.[22]

You really don't need to dig in beyond the abstract of the Losey et al. paper to understand where the controversy came from. The authors summarize a few key points: 1) the Bt toxin is present in corn pollen, 2) corn pollen is dispersed over an area of sixty meters, and 3) monarch butterflies "reared on milkweed leaves dusted with pollen from Bt corn ate less, grew more slowly and suffered higher mortality rates."[166] Seems pretty cut and dry, and that was the sentiment from green organizations and media outlets around the world. Newspaper headlines following the publication of this paper read: "Gene Spliced Corn Imperils Butterflies" (San Francisco Chronicle); "Butterfly Deaths Linked to Altered Corn" (The Globe); and "Genetically Engineered Corn May Have Adverse Effects on Monarch Butterflies" (Los Angeles Times). The word was out, and the agricultural industry was in the hot seat. As Margaret Mellon, from the Washington, D.C.-based Union of Concerned Scientists, stated very pointedly, "Once we heard about the Nature paper, we called reporters and sent out press releases for days. We worked very hard to make this a high-profile issue."[23]

There were, however, issues with the research that provoked several subsequent research projects and risk assessments to investigate, and thereby confirm or refute, the findings of the initial study. The initial publication had three fundamental problems: 1) pollen doses used were not quantitatively measured, 2) pollen from only one type of Bt corn plant was used and subsequently extrapolated to others, and 3) the results provided no basis for the actual risk Bt corn posed to butterflies. This may seem very technical in nature, but it is important to understand the validity of scientific findings, especially in agriculture and food production due to the growing chasm in the consumer base that we have repeatedly highlighted in this book.

So, let's walk through these three issues and provide some follow-up research that was conducted to understand whether we should be concerned with this objection or not. The first problem was with the quantitative approach to applying corn pollen. According to the original article in *Nature*, "Pollen . . . was applied by gently tapping a spatula of pollen over milkweed . . . leaves that had been lightly misted with water. Pollen density was set to **visually match densities on milkweed leaves collected from corn fields** [emphasis added]."[166] Now, I don't know if any of you have looked at corn pollen, but if so, it means that you likely had access to a microscope because the pollen grains are just a bit larger than the limits of what the human eye can distinguish. We are able to distinguish objects 40 microns in size with our naked eyes,[24] and corn pollen comes in at a range of 22-122 microns in diameter.[25] To put it into perspective, an average human hair is roughly 70-100 microns in diameter.[26]

Let's get back to the study in question. The researchers basically "eyeballed" the amount of corn pollen that they were applying to the milkweed leaves. You may be thinking, "What's the big deal? If it isn't harmful to monarchs, then it shouldn't matter how much pollen you expose them to." But this would be a fallacy by way of completely disregarding the principles of toxicology. Remember, the dose makes the poison. If we consume too much caffeine or vitamin A, we could become sick and die. However, at the proper doses these chemicals are completely benign and, in some cases, useful. Research has been conducted to determine how much corn pollen is typically present on milkweed leaves in a corn field, and that number is 171 grains/cm² on average. And that is *within* the corn field where milkweed plants are likely removed by herbicide applications. There is a five-fold reduction in the concentration of corn pollen when you move even two or three meters from the edge of the field.[27]

Let's put these numbers together to paint a picture that drives home the point. We will use 169 grains/cm² instead of 171 as our average amount of corn pollen so we can think of it as a square, thirteen pollen grains down by thirteen pollen grains across. Since the average size of a pollen grain is ninety microns in diameter, this grouping of pollen would measure 1,170 microns by 1,170 microns, or 0.117 cm by 0.117 cm. The area covered by the pollen grains would therefore be 0.0137 cm² for every square centimeter of leaf, or 1.37% of the total leaf area. How confident are you that the researchers were able to "[tap] a spatula of pollen" to get close to the amount typically found on milkweed plants? And again, this is the concentration that could be expected within a corn field, even though in a commercial setting there are likely no milkweed plants present because the weeds are managed. The concentration should be reduced five-fold to represent the amount of corn pollen present on milkweed on the field edge.

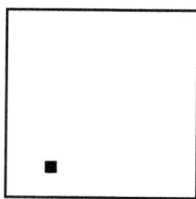

Visual representation to show the average pollen grain concentration on a leaf. According to the Materials and Methods section of the Losey paper, the researchers attempted to achieve this precise concentration by "gently tapping a spatula of pollen over milkweed" that had been misted with water. The open square is 1in x 1in and the blue square is 0.046in x 0.046in.

The second problem with the study was that they only used pollen from one corn hybrid in the trials. This is an issue because different hybrids have different Bt events, which may pose distinct risks to different insect species. In follow-up studies to the Losey paper, six studies determined that there was little risk to monarch larvae from the two most prevalent types of Bt corn. One type of corn that contained the Bt Event 176 was shown to be capable of harming monarch larvae, but this corn never gained considerable market share commercially because the Bt toxin was not expressed at high levels in the stalk (possibly why it was expressed at higher levels in the pollen).

The corn type that Losey et al. used was also tested and was determined to have the next-highest concentration of toxin in the pollen, after Event 176, but no mortality was observed at even 1000 pollen grains/cm². The researchers documented that the highest amount of pollen they measured in the field was 1500 grains/cm², and even if Bt corn was toxic at 1000 grains/cm², which it is not, less than 0.007% of monarch larvae would have encountered this concentration of pollen.[28]

That last statement addresses the final problem with the research—the paper provided no basis for the risk to monarch butterflies. In a lab setting, coating milkweed leaves with pollen from one corn type at levels that are unrealistic to occur in corn fields, let alone on the edges or away from corn fields, may have caused some mortality in monarch larvae. However, as was noted in the last paragraph, the likelihood of monarch butterflies being detrimentally affected in real-world scenarios is exceedingly small. Although the overwhelming body of evidence has now concluded that Bt corn is safe for monarchs, much of the damage to the technology was done from the initial and premature outrage. To this day there are still environmental groups that believe we are killing the monarchs with "deadly" Bt corn.

— GMOs Go from Productivity Focus to Value Add —

The value that GMOs have delivered to agriculture through herbicide- and insect-tolerant crops is most obvious from a production standpoint. While these advancements and innovations ultimately benefit consumers by ensuring an adequate and affordable food supply, this indirect benefit is often overlooked by the casual observer. The next category of GMOs—what we will describe as "value added"—are more overt in their direct benefit to the everyday consumer. But that does not mean the benefits have been welcomed with open arms, as we will see in this section.

Arguably, the most well-known example of a GMO developed to improve the quality of food is "Golden Rice." The term comes from its unique golden color, a stark contrast to the white kernels we usually associate with the grain and is the visual proof of the modifications researchers were able to accomplish. After nine years of research and development, scientists developed rice that contained β-carotene in its kernels. Why was this development worthy of the *Time Magazine* cover in 2000 titled "This rice could save a million kids a year"? Read on to find out.

The carotenoid β-carotene is a pigment that is converted into vitamin A in humans. Vitamin A is important for numerous functions in the body, ranging from the immune system to reproduction, but one of the most notable is its effect on vision. Deficiencies in vitamin A will often appear first as xerophthalmia, an eye disease characterized by abnormally dry eyes and tear ducts that causes night blindness. However, severe and prolonged deficiencies will also increase the incidence of diseases—particularly respiratory and diarrheal—that result in increased mortality rates.

You may be wondering if this is that big of an issue. If you live in the United States, you may have never heard of anyone suffering from night blindness caused by a vitamin A deficiency, let alone dying from lack of the essential vitamin. That is because most people, particularly in developed countries, receive adequate amounts of vitamin A from the foods we eat every day. Common sources are dairy, fish, fortified cereals, and numerous vegetables like carrots, broccoli, sweet potatoes, and squash. Your parents may have told you to eat your carrots as a kid if you wanted to have good eyesight . . . and they were partially right. A half-cup of raw carrots will provide roughly 51% of your daily recommended vitamin A, so eating carrots will certainly contribute to good eye health.[29] But we are fortunate in the developed world to have numerous other sources for vitamin A. If I would have known better as a kid, I could have replied to my mother that I would just eat a slice of pumpkin pie instead, which delivers 54% of my daily value!

Vitamin A deficiency primarily impacts developing countries that have limited access to diverse food options and rely primarily on rice and other carbohydrate-based sources for calories. Pregnant women and children are primarily affected as their higher nutritional demands exacerbate the low levels of this critical vitamin. But children are undoubtedly at the highest risk. It is estimated that nearly one-third of all children in the world under the age of five suffer from vitamin A deficiency, which is the leading cause of childhood blindness.[30] Further, this deficiency is responsible for approximately 2% of all deaths in this age group globally.[31] According to statistics from the World Health Organization, approximately 250 million children under the age of five were vitamin A deficient in 2012.[32] Providing these children with sources of vitamin A, they claim, could prevent between 1.3 to 2.5 million deaths.

It was these sobering statistics that led researchers to begin working on solutions through genetic engineering. Because rice is the primary component of much of the developing world's diet, it was an ideal target for improvement. Rice naturally contains β-carotene, but only in its leaves and stems, so researchers began to work on finding genes that would turn on β-carotene production in the kernels. The man behind the creation of Golden Rice—or at least one of the men and women who worked on the project, and the one who graced the cover of July 2000 issue of *Time Magazine*—was Ingo Potrykus.

If you want to learn firsthand about the long and arduous journey to create Golden Rice, I encourage you to read Potrykus' blog titled "The 'Golden Rice' Tale" found on agbioworld.org.[33] He outlines how he got his start in genetic engineering, the numerous people who were involved in the project, and the breakthroughs they made along the way. Creating Golden Rice required the introduction of several genes because they essentially needed to augment an entire metabolic pathway, but interestingly, the critical gene that produced the β-carotene was isolated from daffodils. In what seems more like the fairy tale ending of a movie than real life, Potrykus explains how the first successful Golden Rice seeds produced were presented to him at his farewell symposium, an event he describes as occurring "on 31 March 1999, the date I had to retire because I had passed the age limit." In addition to the gift, his colleague Xudong Ye presented the successful results of the Golden Rice project to the public for the first time.

Potrykus' passion for his work is infectious. He describes the scientific breakthrough discovered by him and his team, highlighting how it would be used to "solve an urgent need and to provide a clear benefit to the consumer, and especially to the poor and disadvantaged." But however significant the success, the real work began when they tried to introduce it to the world. The unwavering commitment of those involved with the Golden Rice project was to make the technology free of charge to subsistence farmers in developing countries. But Potrykus and his colleagues realized very quickly the hurdles involved with licenses, patent applications, intellectual property rights, technical property rights, and especially technology transfer to developing countries.

Ultimately, with the help of both public and private cooperators, they navigated the complex waters of international licensing and put in place a system to perform all the necessary safety assessments, as well as socio-economic and environmental impact studies. Their cooperator from the private sector was Zeneca, who received exclusive rights for commercial use of the seed. Before you jump to conclusions, Zeneca agreed that they would in turn support the humanitarian use of any current or future technologies developed from this project. This meant that the farmers Potrykus and his colleagues were trying to support would be ensured free access to the technology. The company would be able to sell the rice commercially in the developed world to pay for the investment, but is that a bad thing? Additionally, the technology would only be transferred to institutions that would use conventional breeding to introduce the trait into the best, locally adapted lines.

All good? Not quite. GMO opposition, as it turns out, takes precedence over humanitarian good. Potrykus concludes his "tale" of Golden Rice by first bulleting numerous facts about the project, which I will provide in full:

Golden Rice" fulfills all the wishes the GMO opposition had earlier expressed in their criticism of the use of the technology, and it thus nullifies all the arguments against genetic engineering with plants in this specific example.

- Golden Rice has not been developed by and for industry.

- It fulfills an urgent need by complementing traditional interventions.

- It presents a sustainable, cost-free solution, not requiring other resources.

- It avoids the unfortunate negative side effects of the Green Revolution.

- Industry does not benefit from it.

- Those who benefit are the poor and disadvantaged.

- It is given free of charge and restrictions to subsistence farmers.

- It does not create any new dependencies.

- It will be grown without any additional inputs.

- It does not create advantages to rich landowners.

- It can be resown every year from the saved harvest.

- It does not reduce agricultural biodiversity.

- It does not affect natural biodiversity.

- There is, so far, no conceptual negative effect on the environment.

- There is, so far, no conceivable risk to consumer health.

- It was not possible to develop the trait with traditional methods, etc.

This list was not enough for those bound and determined to stop GMOs. Anti-GMO activists, like Vandana Shiva, who we will hear more from in later sections, began their propaganda efforts to discredit the technology, which continues to this day. Greenpeace issued a statement in 2001 claiming that adults would need to eat twenty pounds of cooked Golden Rice every day to prevent vitamin A deficiency, attempting to cast doubt on the benefit the technology would provide.[34] The only problem with this claim was that the bioavailability of β-carotene from Golden Rice was not known at the time, so it was pure propaganda.

Potrykus, ever the diplomat, responded to the Greenpeace criticism by first acknowledging that he could find common ground with them over their opposition to the large PR campaigns of agricultural companies surrounding Golden Rice.[35] It turns out that, in addition to eventually being able to sell the technology in developed countries, another benefit that accrued to the corporations involved in Golden Rice was the goodwill they could generate from this humanitarian effort in the eye of the public. (Again, not saying this is a bad thing, especially if it gets the needed technology in the hands, or mouths, of those that need it). However, Potrykus went on to point out that the effectiveness of Golden Rice towards its mission would not be fully known until the trait was bred into commercial varieties, until bioavailability studies were completed, and until nutritional studies were conducted with individuals suffering from vitamin A deficiency.

Additionally, he noted that Greenpeace was using Recommended Daily Values (RDVs) in their calculations, but to mitigate the symptoms of vitamin A deficiency, there was the general consensus of the medical community that the amount needed is significantly lower. Even if the RDVs were the amount needed, which again is likely far more than what is needed to prevent deficiency, preliminary data would suggest an adult would need to eat 1.7 to 3.3 pounds daily, not 20. Finally, he made probably the most insightful observation: his team had only just created the first experimental prototype and were already in the range of 20-40% of the daily allowance that Greenpeace is arguing for. Therefore, Golden Rice was certainly a viable complementation to the mitigation efforts aimed to help undernourished people in developing countries and should continue to be researched and developed.

Development did in fact continue and Golden Rice 2 was launched in 2005. Researchers replaced the daffodil gene with one from corn, which resulted in 23 times the amount of β-carotene as the first-generation lines. Now, only a cup of cooked rice was needed to supply adults with half of their required vitamin A. Field trials were also being conducted and cooperation with numerous countries was underway to develop locally suitable varieties. Today, breeding programs have developed varieties that are able to deliver 80 to 110% of the recommended daily intake of vitamin A for children and women.[36]

So, is Golden Rice fulfilling its promise to help save the lives of children in developing countries? We are, after all, over twenty years post-discovery. The answer is a resounding no. Not because it has been unsuccessful, but because it has been untried. Golden Rice continues to battle political and regulatory headwinds in its fight for approval, like the Cartagena Protocol on Biosafety and the concept of the precautionary principle. The Cartegena Protocol is critical in our discussion around GMO adoption and opposition, and we will devote time to the agreement in a subsequent section. In this context, it is most important to define the precautionary principle. According to the Protocol:

"In order to protect the environment, the precautionary approach shall be widely applied by States according to their capabilities. Where there are threats of serious or irreversible damage, lack of full scientific certainty shall not be used as a reason for postponing cost-effective measures to prevent environmental degradation."[37]

For Golden Rice, the inability to prove "full scientific certainty" that there would be no harm to the environment has overridden the ability to provide a solution for children in need.

In late 2019, government regulators in the Philippines approved Golden Rice to be used in food, feed, and in processed form, declaring it "as safe as conventional" varieties of rice.[38] After taking nearly a year to progress through the public comment period, an application was submitted in October of 2020 to allow breeders to propagate individual varieties, each of which needs to be tested and registered prior to commercial release. The application appeared to be "stuck due to the reluctance of the Department of Environment," according to Muhammad Abdur Razzaque, the Philippine Agriculture Minister.[39] However, on July 21, 2021, the Philippines became the first country in the world to approve the commercial production of Golden Rice. Although it feels like the finish line is in sight, it is hard not to be haunted by the words of senior science research specialist, Reynante Ordonio, from the Philippine Rice Research Institute, who states, "2023 is our estimate. Personally, I've learned not to expect much but we remain hopeful."[40]

1 Act of May 20, 1862 (Homestead Act), Public Law 37-64, 05/20/1862; Record Group 11; General Records of the United States Government; National Archives. https://www.ourdocuments.gov/doc.php?flash=false&doc=31

2 History.com Editors (2021). *Homestead Act.* History. https://www.history.com/topics/american-civil-war/homestead-act

3 Tippett R (2015). *Agriculture and food statistics: USDA charts the essentials.* Carolina Demography. https://www.ncdemography.org/2015/05/11/agriculture-and-food-statistics-usda-charts-the-essentials/

4 USDA NASS (2019). *United States Summary and State Data.* 2017 Census of Agriculture Geographic Area Series 1(51). United States Department of Agriculture National Agricultural Statistics Service. AC-17-A-51. https://www.nass.usda.gov/Publications/AgCensus/2017/Full_Report/Volume_1,_Chapter_1_US/usv1.pdf

5 Top Ag News (2019). *Farmers report how much time it takes to plant their corn crop.* ProAg. https://www.proag.com/news/farmers-report-how-much-time-it-takes-to-plant-their-corn-crop/

6 Funke T, Han H, Healy-Fried ML, Fischer M, and Schönbrunn E (2006). Molecular basis for the herbicide resistance of Roundup Ready crops. *PNAS 103*(35), 13010-13015. https://doi.org/10.1073/pnas.0603638103

7 USDA ERS (2020). *Recent trends in GE adoption.* United States Department of Agriculture Economic Research Service. https://www.ers.usda.gov/data-products/adoption-of-genetically-engineered-crops-in-the-us/recent-trends-in-ge-adoption.aspx

8 Michigan Farm Bureau (2016). *Biggest irrigated crop? The American lawn.* Michigan Farm News. https://www.michiganfarmnews.com/biggest-irrigated-crop-the-american-lawn

9 Dille JA, Sikkema PH, Everman WJ, Davis VM, and Burke IC (2015). Perspectives on corn yield losses due to weeds in North America. Poster presented at: 2015 Weed Science Society of America Conference; February 9, 2015; Lexington, KY. https://wssa.net/wp-content/uploads/WSSA-2015-Corn-Yield-Loss-poster-updated-calc.pdf

10 Krupke CH, Bledsoe LW, and Obermeyer JL (2010). *European corn borer in field corn.* Field Crops. Purdue Extension. E-17-W. https://extension.entm.purdue.edu/publications/E-17.pdf

11 Hellmich RL and Hellmich KA (2012). Use and impact of Bt maize. *Nature Education Knowledge 3*(10), 4. https://www.nature.com/scitable/knowledge/library/use-and-impact-of-bt-maize-46975413/

12 Ostlie KR, Hutchison WD, and Hellmich RL (1997). Bt corn and European corn borer. *Faculty Publications: Department of Entomology.* 597. https://digitalcommons.unl.edu/entomologyfacpub/597

13 Cullen EM, Gray ME, Gassmann AJ, and Hibbard BE (2013). Resistance to Bt corn by western corn rootworm (Coleoptera: Chrysomelidae) in the U.S. Corn Belt. *Journal of Integrated Pest Management 4*(3), D1-D6. https://doi.org/10.1603/IPM13012

14 USDA ERS, Recent trends in GE adoption.

15 Romeis J, Naranjo SE, Meissle M, and Shelton AM (2019). Genetically engineered crops help support conservation biological control. *Biological Control 130*, 136-154. https://doi.org/10.1016/j.biocontrol.2018.10.001

16 Moazamian E, Bahador N, Azarpira N, and Rasouli M (2018). Anti-cancer parasporin toxins of new *Bacillus thuringiensis* against human colon (HCT-116) and blood (CCRF-CEM) cancer cell lines. *Current Microbiology 75*(8), 1090–1098. https://doi.org/10.1007/s00284-018-1479-z

17 UCSD (n.d.). *Bacillus thuringiensis: History of Bt.* University of California San Diego. http://www.bt.ucsd.edu/bt_history.html

18 Watkins PR, Huesing JE, Margam V, Murdock LL, and Higgins TJV (2012). Insects, nematodes, and other pests. In Altman A & Hasegawa PM (Eds.), *Plant biotechnology and agriculture* (353-370). Academic Press. https://doi.org/10.1016/B978-0-12-381466-1.00023-7

19 Mahr S (2018). *Bacillus thuringiensis.* Wisconsin Master Gardener. https://walworth.extension.wisc.edu/files/2018/06/BT-Insecticide-Article.pdf

20 Fernandez-Cornejo J, Nehring R, Osteen C, Wechsler S, Martin A, and Vialou A (2014). Pesticide use in U.S. agriculture: 21 selected crops, 1960-2008. *Economic Information Bulletin 124.* United States Department of Agriculture Economic Research Service. https://www.ers.usda.gov/webdocs/publications/43854/46734_eib124.pdf

21 Baden-Mayer A (2012). *Stop GMO Bt corn before it kills again!* Organic Consumers Association. https://www.organicconsumers.org/essays/stop-gmo-bt-corn-it-kills-again

22 Losey JE, Rayor LS, and Carter ME (1999). Transgenic pollen harms monarch larvae. *Nature 399*(214). https://doi.org/10.1038/20338

23 Pew Trusts (2002). *Three years later: Genetically engineered corn and the monarch butterfly controversy.* Pew Initiative on Food and Biotechnology. https://www.pewtrusts.org/~/media/legacy/uploadedfiles/wwwpewtrustsorg/reports/food_and_biotechnology/vfbiotechmonarchpdf.pdf-

24 Wong Y (n.d.). *How small can the naked eye see?* Science Focus. https://www.sciencefocus.com/the-human-body/how-small-can-the-naked-eye-see/

25 IMS Health Incorporated (2021). *Corn (Zea).* PollenLibrary.com. https://www.pollenlibrary.com/Genus/Zea/

26 Clearstream (2021). *Micron size comparison chart.* Clearstream Filters Inc. https://www.clearstream.ca/micron-size-comparison-chart

27 Sears MK, Hellmich RL, Stanley-Horn DE, Oberhauser KS, Pleasants JM, Mattila HR, Siegfried BD, and Dively GP (2001). Impact of Bt corn pollen on monarch butterfly populations: A risk assessment. *PNAS 98*(21), 11937-11942. https://doi.org/10.1073/pnas.211329998

28 Pew Trusts, Three years later.

29 ODS (2021). *Vitamin A: Fact sheet for health professionals.* National Institutes of Health Office of Dietary Supplements. https://ods.od.nih.gov/factsheets/VitaminA-HealthProfessional/

30 Gearing ME (2015). *Good as gold: Can Golden Rice and other biofortified crops prevent malnutrition?* Science in the News, Harvard University. https://sitn.hms.harvard.edu/flash/2015/good-as-gold-can-golden-rice-and-other-biofortified-crops-prevent-malnutrition/?web=1&wdLOR=c544D7A75-D3E9-458B-BD94-8679C06EBC9F

31 Wirth JP, Petry N, Tanumihardjo SA, Rogers LM, McLean E, Greig A, Garrett GS, Klemm RD, and Rohner F (2017). Vitamin A supplementation programs and country-level evidence of Vitamin A Deficiency. *Nutrients 9*(3):190. doi: 10.3390/nu9030190.

32 Mayer J (2021). *Vitamin A Deficiency-Related Disorders (VADD).* Golden Rice Project. Golden Rice Humanitarian Board. http://www.goldenrice.org/Content3-Why/why1_vad.php

33 Potrykus I (2011). *The 'Golden Rice' tale.* AgBioWorld. http://www.agbioworld.org/biotech-info/topics/goldenrice/tale.html

34 Regis E (2019). The true story of the genetically modified superfood that almost saved millions: The imperiled birth – and slow decline – of Golden Rice. Foreign Policy. https://foreignpolicy.com/2019/10/17/golden-rice-genetically-modified-superfood-almost-saved-millions/

35 Potrykus I (2001). *Potrykus responds to Greenpeace criticism of 'Golden Rice.'* AgBioWorld. http://www.agbioworld.org/biotech-info/topics/goldenrice/criticism.html

36 Mayer J (2021). *Golden Rice Project.* Golden Rice Humanitarian Board. http://www.goldenrice.org/

37 Eggers B & Mackenzie R (2000). The Cartegena protocol on biosafety. Journal of International Economic Law 3(3):525-543. https://doi.org/10.1093/jiel/3.3.525

38 Rees M and Harvey P (2021). *Golden Rice could fight deadly vitamin A deficiency now. Why do farmers have to wait another 3 years to grow it?* Genetic Literacy Project. https://geneticliteracyproject.org/2021/01/06/golden-rice-could-fight-deadly-vitamin-a-deficiency-now-why-do-farmers-have-to-wait-another-3-years-to-grow-it/

39 Golden Rice Project (2020). *Golden Rice release stuck due to DoE reluctance: Minister.* The Daily Observer. http://www.goldenrice.org/PDFs/Golden%20Rice%20stuck%20-%20Daily%20Observer%202020.pdf

40 Simeon LM (2020). *Golden Rice to be available in 3 years.* The Philippine Star. https://www.philstar.com/business/2020/11/24/2058936/golden-rice-be-available-3-years

Non-GMO Verified:
Opposing Forces

— Non-GMO Reaches a Fever Pitch —

From the proliferation of the "Non-GMO Project Verified" food labels to the continued mainstream mantra that GMOs are inherently suspect, it could be argued that our current landscape is hostile to the concept of GMOs. What are your thoughts?

While no empirical data has been put forth to convince the American consumer that non-GMO foods are somehow better for them than their conventional counterparts, the market for foods with this label is growing rapidly. According to FoodChain ID, a technical administrator of the Non-GMO Project Verified program, the label "is the fastest growing label in the natural products industry, representing $25 billion in annual sales."[1] They add that "a growing number of consumers are specifically searching for products carrying the NGP label and that products with the label see measurable increases in sales." Just so we're all on the same page—the reason for getting the label is that consumers have been convinced they need it, so having it will result in increased sales.

And this sales growth is not in the low-price tier; rather, it is a market where products can realize a significant premium. A study that quantified the premiums non-GMO foods were able to demand over their conventional counterparts found food categories ranged from a low of 9.8% for ice cream to a high of 61.8% for salad and cooking oils.[2] It is no wonder food companies would go this route. Increased sales and increased prices? Yes, please.

But there is a simple fact that we should all be aware of when we peruse the grocery store aisles—there are only a few GMO plant types that are commercially available today. There are obviously corn, soybean, and cotton, all of which we discussed in the previous section. These are used in numerous processed foods, oils, and animal feed. Then you have canola and sugar beets, both of which can be herbicide-tolerant and used for oil and sugar, respectively. There is herbicide-tolerant alfalfa, which is fed to cattle. Also, roughly 75% of papaya grown in Hawaii is genetically modified as it was nearly wiped out by a virus in the 1990s, but genetic engineering single-handedly saved the industry. After that, there are just three additional foods—disease-resistant summer squash that is not widely grown, a few varieties of apples that resist browning after being cut, and some varieties of potatoes that are insect and disease resistant as well as resistant to bruising. If you were counting, that was a total of ten GMO foods available.[3] Do you feel like you see the non-GMO sticker on a lot more products than just the ones that would include these food types?

How about an example: Emerald Natural Almonds. They have the non-GMO label. But the only ingredient is almonds, of which no GMO varieties exist. Interesting. A more appropriate word would be misleading, and it is widespread. An article from the Missouri Farm Bureau points this out with the availability of items like non-GMO water, pink Himalayan salt, kitty litter, dish soap, and even condoms![4] To the author's point, none of these items even contain genetic material, let alone genetically modified versions of it. They continue to point out the premium that this labeling affords companies, with Amazon selling a ten-pound bag of The Good Earth Non-GMO Project Verified clumping cat litter for $18.99, while ten pounds of your basic Arm & Hammer clumping cat litter will only run $5.30 at Walmart. Maybe we should ask what that $13.69 price premium is getting us in that top-shelf kitty litter.

food for thought...

Papayas were "single-handedly saved" by GMO technology? That sounds like a pretty bold statement, but in light of the risk posed to the tree by the papaya ringspot virus, the lack of other remedies, and the concentration of the papayas to a specific area of Hawaii, I'd say it is accurate. So how did a researcher at a university

food for thought... (cont'd)

in upstate New York pull off this feat?

Hawaii native Dennis Gonsalves began his career at Cornell University as a phytopathologist in the late 1970s and became aware of the challenge facing papaya producers during a trip he made home to the Big Island of Hawaii. A virus was causing significant damage to papaya trees in the state, and was rapidly encroaching on the area of the Big Island where over 90% of the papayas were grown.

After his trip home, Gonsalves directed his research towards finding a solution for papaya farmers struggling with the ringspot virus. He began working on the project in 1985, and by 1991, he had successfully introduced a gene into the papaya plant that would convey resistance to the virus. Remember, this would be three years before the FLAVR SAVR tomato made it to store shelves and five years before herbicide tolerant crops were introduced. He was on the cutting edge!

Gonsalves approached the solution similar to that of a vaccine. He was able to isolate the gene of the viral coat protein and use a "gene gun" to introduce the gene into the plant. (I know we didn't talk about this method for creating GMOs, but all you need to know is that small metal particles are covered with the desirable DNA and "shot" into the plant, hoping that the DNA makes it to the nucleus and is incorporated into the plant's genome.)

Once the first transgenic papaya plants were created in the lab and greenhouse, field testing was immediately initiated in 1992. This happened to be the same year that the virus made it into the main papaya-producing area on the Big Island. By 1994, the virus was widespread through the region and by 1998, the virus had cut papaya production in half from 1992 levels. Gonsalves and team were faced with...

immense pressure to bring their solution to market, but they still needed to clear regulatory safeguards. They submitted the application to commercially produce their GMO papayas in 1995 and three years later, they received approval for two virus-resistant varieties, Rainbow and SunUp and they provided Hawaiian growers the papaya seeds at cost.[6]

I would encourage all of you to watch the YouTube video titled *The Story of Rainbow Papaya – Why Public Sector Biotechnology Matters.*[7] You get to hear the story straight from Dennis Gonsalves himself, and it is hard not to be inspired to make a difference in the world after listening to his passion. He has a line in the video that I think is extremely relevant to our discussions through this section "When you put the human part of biotechnology into the equation, then it's easier… to continue the work because you are trying to help people. Too many people think it's just a technology." Yes, there are people at the center of these biotechnology solutions that are being brought to market, and real people are going to benefit from the scientific advancements.

This should be a sobering thought as we continue to look at how the anti-GMO activists reacted to the new papaya varieties. Gonsalves and others next turned their attention next to Thailand, a papaya producing country that was also being impacted by the virus. However, the activists located and damaged the fields where the papaya trials were being conducted; farmers participating in the trials were targeted and their trees were destroyed. Ultimately, the public backlash was too much for the Thai government and they shut down the program. Papaya farmers in that country are still struggling with the impacts of being deprived the available tools that could help them manage this disease, made inaccessible simply because of peoples' intolerant views on a technology.

What about the orange juice aisle? Tropicana. Simply Orange. Yep, the Non-GMO label is on both of those bottles. Yet, these bottles contain only the juice of oranges, of which no genetically modified varieties exist. This example is of particular concern because the citrus industry—specifically in Florida—is under attack from a disease that is threatening orange production as we know it. Citrus huanglongbing (HLB), also known as citrus greening, is a disease that can cause stunted growth, yellow leaves, and other symptoms on the tree, but the primary damage is done to the fruit. Trees will produce small, oblong, off-colored oranges that, most importantly, are low in soluble solids and high in acids, resulting in bitter juice.

HLB is caused by a bacteria that is vectored by a small insect called the Asian citrus psyllid. After only fifteen to thirty minutes of feeding on infected leaf tissue, the psyllid can pick up the bacteria and transmit the disease for the rest of its life. Additionally, an infective psyllid can transmit the disease to new trees after feeding on them for as little as fifteen minutes.[7]

The Asian citrus psyllid and HLB are native to the Punjab region of Pakistan and India, but both have spread into the rest of the Indian subcontinent, Africa, the Saudi Arabian Peninsula, southeast Asia, Belize, Cuba, Mexico, and the southeastern United States.[8] While it has been identified in California, strict quarantine protocols have limited its spread through much of the state. The biggest impact in the United States has been in Florida where 95% of all oranges grown in the state are processed for juice. From 2005, when the disease was first identified, to 2018, orange tree acreage decreased by 37% and yields decreased by 81%, a drop from 242 million to 45 million boxes of oranges—due almost entirely to this disease![9]

So, why is the "Non-GMO" label for orange juice of particular concern, aside from the obvious reason that no available orange is genetically modified? It is because of the devastation HLB is causing and the current lack of options available to control this epidemic. At this point, it would be madness to turn down any possible solution, regardless of the source. Producers and researchers are working tirelessly to find best management practices, but because of the complex relationship between the host orange tree, the bacterial pathogen, and the insect vector, it is exceedingly difficult. There is no cure for the disease, and current management efforts can only slow the decline of orange trees through irrigation and nutrient management practices. Unfortunately, the best course of action for many orange farmers has been to simply stop farming.

Various other mitigation efforts are being tested, like spraying antibiotics (since the disease is caused by a bacteria) and even releasing predatory wasps that feed on the Asian citrus psyllids. While both methods can have some effect (to varying degrees), neither provides a reliable method to control the disease. One promising idea was announced in 2020, with expectations that it could be available to growers within three years. Hailing Jin, a researcher at the University of California, Riverside, isolated genes from Australian finger limes, which are tolerant to citrus greening, and identified a peptide present in the fruit that appears to control the bacteria when applied foliarly to orange trees.[10] The peptide is more stable than antibacterial sprays, and initial studies have shown it to be effective with relatively few applications per year.

The development of a peptide that is present in tolerant citrus varieties is the closest the industry has come to a potential solution for the citrus greening epidemic. It has also been discussed that the peptide could be developed into a sort of vaccine that could be injected into young orange trees, providing season-long protection by inducing the plant's innate immune system. Yet another possibility could be the integration of the gene that encodes for this peptide directly into the plant's genome. Instead of isolating the gene from another plant, synthesizing and manufacturing the resulting peptide, and then spraying the peptide onto the plant (or potentially injecting it into the plant), could we just have the plant produce the peptide itself? This would be the genetic engineering route, and one that is also being investigated for possible solutions. Is one better or worse than the other? Aside from one option preventing the oranges and orange juice from being labeled non-GMO and the other allowing the all-important butterfly to remain on the bottle, what is the difference? Before we use virtue signaling as a marketing gimmick to drive sales, it may be worth looking at the situation honestly to determine if a GMO, or even non-organic practice, could save an entire industry.

This premise brings us full circle to the original contention with the Non-GMO Project (NGP) initiative and the non-GMO movement in general. We hear frequently these days that we should "follow the science." If we take that statement as an actual request, where *does* the science lead us? According to the NGP website, we would believe first that "there is no scientific consensus on the safety of GMOs."[11] Sounds scientific. I mean, they do use the word *scientific* in the statement. But according to the American Association for the Advancement of Science in their 2012 *Statement by the AAAS Board of Directors on Labeling of Genetically Modified Foods*, "Efforts to require labeling of foods containing products derived from genetically modified crop plants . . . are not driven by evidence that GM foods are actually dangerous." In fact, they state that "the science is quite clear: crop improvement by modern molecular techniques of biotechnology is safe."[12]

Citing the AAAS again, even research by the European Union stated the conclusion from "more than 130 research projects, covering a period of more than 25 years of research and involving more than 500 independent research groups, is that biotechnology, and in particular GMOs, are not per se more risky than e.g. conventional plant breeding technologies." They broaden the argument by stating:

> *"The World Health Organization, the American Medical Association, the U.S. National Academy of Sciences, the British Royal Society, and every other respected organization that has examined the evidence has come to the same conclusion: consuming foods containing ingredients derived from GM crops is no riskier than consuming the same foods containing ingredients from crop plants modified by conventional plant improvement techniques."[13]*

If we are to follow the science, should we believe the organization focused on marketing premium products and driving sales, or the American Association for the Advancement of Science and the other scientific organizations that have all come to the same conclusion?

food for thought...

I told you we would save our discussion on the COVID-19 vaccines for later, and I think this is the perfect place. The concerted effort by the public and private sector to drive research, development, and initial deployment of a vaccine in less than a year was previously thought impossible. But as of this writing, we now have three vaccines that are authorized and recommended for use by the CDC in the United States: the Pfizer-BioNTech, Moderna, and Johnson & Johnson vaccines, with two others available globally. I am just going to come out and say it, if this doesn't put an end to the anti-GMO debate, nothing will.

Each of the three approved vaccines in the US are examples of groundbreaking technologies. And I mean that in the most literal sense as they are the first vaccines of their kind to be approved for human use. The Pfizer and Moderna vaccines are both mRNA vaccines and the J&J vaccine is a non-replicating viral vector vaccine, which I will state again, are the first examples of these technologies to ever be approved for humans. So how do they work?

To understand the mechanism of the mRNA vaccines, we need to understand the "central dogma of molecular biology." In brief," DNA in the nucleus of a cell is copied into mRNA (transcription) that is transported out of the nucleus and into the cytoplasm where ribosomes will synthesize proteins (translation). The two mRNA vaccines in question provide that second piece of the central dogma. This piece of genetic information is the transcribed sequence of the coronavirus gene that codes for the spike protein of the virus. It does not interact with our DNA because it does not enter the nucleus of the cell, but it is recognized by our cells as mRNA that needs to be translated. In this way, the viral genetic material that is floating in the cytoplasm essentially "tricks" our cells to synthesize the spike protein of the virus, which accumulates on the surface of the cell. Our immune system recognizes that the spike protein is a foreign body and responds by destroying the foreign invaders, creating antibodies to block the virus from being able to infect more cells, and essentially recruiting "memory" cells that will recognize any future infections...

While no recombinant DNA technology is used in this process, genetic material from another organism is injected into our bodies, which causes our cells to produce virus proteins. But we buy non-GMO orange juice even though there are no genetically modified oranges? I guess it is okay because the CDC's website makes the statement, "COVID-19 mRNA vaccines give instructions for our cells to make a **harmless piece** of what is called the 'spike protein.'" [Emphasis in original]

How about the J&J vaccine that uses a non-replicating viral vector to deliver immunity? A non-infective virus, an adenovirus, in this case, is genetically modified to contain the gene coding for the spike protein of the coronavirus. This genetically modified virus is then injected into our bodies where it makes its way into our cells. At this point, the same mechanism that was just described for mRNA vaccines is carried out to enable immunity to future infections. This type of vaccine is nearly identical to the common practice of using weaker versions of infective viruses to trigger an immune response, except for the fact that they use a genetically modified carrier virus that only contains genes for certain proteins of the target virus.

The CDC provides comforting language to describe this type of vaccine as well, outlining the first step of the process by stating, "the vector (**not** the virus that causes COVID-19, but a different, harmless virus) will enter a cell in our body and then use the cell's machinery to produce a **harmless** piece of the virus that causes COVID-19." [Emphasis in original] Using the word "harmless" twice in the same sentence? I wonder if they were worried about the potential fear people would have with this vaccine if it were honestly reported to people that they were directly injecting GMOs into their own cells? .●

— The Right to Own . . . Life? —

Can we own a patent on life? Talk about a moral dilemma. But that is the final discussion point we need to unpack on the GMO argument that faces us today. Yes, you can patent life, in the context of a new plant species or of a genetically engineered organism, and both are permissible under the law. How should we feel about this?

The first legal protection provided for plant breeders was passed by Congress in 1930 with the Plant Patent Act (PPA). The legislation provided patent protection for seventeen years to anyone who had discovered or invented a new variety of plant, but it was limited to asexually reproduced plants (meaning without using pollination and seeds) and excluded tuber-propagated plants (those where the root can be used to generate a new plant, like the potato).[14] This was certainly before the advent of genetically modified organisms, but it still granted important rights to conventional plant breeders of the time. Many credit the work of botanists and horticulturists, like Luther Burbank, as the catalyst for the legislation because of the amount of time and effort these scientists devoted to bringing new and improved crop types and varieties to the public. Burbank is credited with introducing over 800 new varieties of plants, including the Shasta daisy, more than 100 varieties of plums, spineless cacti for forage in the desert, and the Russet Burbank potato, the most popular potato in the United States today. In front of Congress, Thomas Edison testified that "nothing that Congress could do to help farming would be of greater value and permanence than to give the plant breeder the same status as the mechanical or chemical inventors now have through patent law. There are but a few plant breeders. This . . . will, I feel sure, give us many [Luther] Burbanks."[15] Burbank, while enjoying a lasting legacy from his plant varieties and serving as the inspiration for Santa Rosa's annual Rose Parade, was awarded sixteen plant patents posthumously as he died four years before the Plant Patent Act of 1930.[16]

You may be wondering about the exclusion of "tuber-propagated" plants from the 1930 Act, especially after our emphasis on the man who created the Russet

Burbank potato, which is in fact a tuber. Enter the Plant Variety Protection Act (PVPA) of 1970. Essentially, this Act expanded the 1930 legislation to include sexually reproduced (or reproduced by seeds) and tuber-propagated plants under patent protection. However, unlike the patent granted by the PPA decades earlier, the PVPA issued a plant variety certificate, essentially prohibiting others from selling, reproducing, importing, exporting, or producing the protected variety. There were also three exemptions included in the legislation, one of which would be extremely important in subsequent amendments and litigation. This exemption permitted farmers to save seeds from protected varieties and then use the saved seeds in the production of a crop, or even sell the saved seeds, all without infringement on the patent.

It wouldn't take long for this exemption to find its way into a legal battle. The Asgrow Seed Company brought a lawsuit against Iowa farmers Denny and Becky Winterboer, a case that was ultimately decided by the Supreme Court in January 1995. The Winterboers planted 265 acres of two Asgrow soybean varieties and subsequently sold the entire harvested crop as seeds to other farmers, an amount determined to be enough to plant 10,000 acres. The varieties were being sold for more than $16.00 per bushel by Asgrow, but the farmers were selling the harvested crop at a price of $8.70 per bushel.[17]

Upon learning of this, Asgrow employed a local farmer to purchase seed from the Winterboers, who stated to the undercover farmer that they were selling soybean seed that was "just like" the Asgrow varieties. This is what drove Asgrow's argument when they brought a lawsuit against the Winterboers, and why the US Supreme Court eventually ruled in favor of Asgrow in an 8-1 decision with the opinion delivered by Justice Antonin Scalia. The court ruled that although the PVPA allows a farmer to save seed to replant his fields, and then to sell the saved seed, the statute "prohibits growing protected seed as a 'step in marketing' it as seed for planting." According to the court, the Winterboers planted and harvested the Asgrow seed solely to market a protected seed variety, and therefore forfeited their eligibility of protection under the PVPA exemption.

This lawsuit drove the 1994 amendment to the PVPA to prohibit the sale of all saved seed by a farmer without the permission of the variety owner, as well as extending the patent protection from seventeen to twenty years. It was simultaneously a big win for the seed companies and a black eye for the industry, as it was perceived as big ag coming down on small farmers. Referred to as a "David and Goliath" struggle by news publications, big ag was cast as the mega-corporation simply looking to squeeze the little guy for all they were worth.

We jumped slightly out of chronological order, but you may remember that earlier in the book we briefly mentioned the first GMO that was patented in 1980. Let's return to this briefly to establish some precedence around patenting living things. In this case, General Electric employee Ananda Chakrabarty developed a new bacterium from the *Pseudomonas* genus that broke down crude oil, a novel solution that could aid in remediating oil spills. Chakrabarty's patent application was initially rejected by the patent examiner and upheld by the Patent Office Board of Appeals, which stated that living things were not patentable. The US Court of Customs and Patent Appeals reversed the decision and the Commissioner of Patents and Trademarks, Sidney Diamond, appealed to the US Supreme Court. The Supreme Court ruled in favor of Chakrabarty in a narrow 5-4 decision under the justification that "while laws of nature, physical phenomena, and abstract ideas are not patentable, . . . a live, human-made micro-organism is patentable subject matter" and "constitutes a 'manufacture' or 'composition of matter' within that statute."[18]

The fact that the material in question was living was irrelevant according to the ruling decision (although it was foundational to the dissenting opinion); the primary concern was that this material was man-made and not existing in nature. While I am not a subject matter expert in law, particularly when it relates to bioethics, I am reminded that the ability to patent and copyright intellectual property and invention has been integral to innovation in the United States. While there are necessary limits in place on what can and can't be patented, from my perspective, the ability to protect discovery has driven great progress in our society and I am grateful for the protections that have allowed great minds to continue to think about what is possible.

Let's jump back to the agricultural side of the law and cover one more landmark case on the subject of patent protection. In 2001, the US Supreme Court ruled on whether utility patents could be issued for plants. Most patent applications that are filed with the US Patent and Trademark Office (USPTO) each year are for utility patents, or those that fall into the categories of process, machine, manufacture, composition of matter, or any new and useful improvement thereof. However, the PPA and PVPA focused on the idea of plant patents and plant protection certificates, essentially protecting the new variety or plant type that was created.

The Pioneer Hi-Bred seed company brought a lawsuit against J.E.M. Ag Supply, Inc. because they bought patented seeds from Pioneer that had license agreements on the bag and then resold them. In the 6-2 Supreme Court ruling in favor of Pioneer, Justice Clarence Thomas stated in his opinion that utility patents could be issued for plants and that "denying patent protection . . . simply because such coverage was thought technologically infeasible in 1930 . . . would be inconsistent with the forward-looking perspective of the utility patent statute."[19]

So, to summarize, plants could be patented under the PPA in 1930, protected plants were expanded to include sexually reproduced and tuber-propagated plants with the PVPA in 1970, the first genetically modified organism was patented in 1980, the PVPA was amended and ruled on in 1994 to prevent the sale of protected plants, and it was ruled that utility patents could be issued for plants in the 2001 ruling. Now that you have the complete backstory, let's discuss an example that has certainly garnered its share of publicity and see where you land on ag companies being able to "patent life."

Percy Schmeiser was a Canadian canola farmer from Saskatchewan who gained notoriety in the late 1990s when he went toe-to-toe with Monsanto over patent infringement claims. Maybe you saw his story documented in the movie *David vs Monsanto* in 2009. Or maybe references to his story in "documentaries" like *The Future of Food* (2004) or *Seeds of Death: Unveiling the Lies of GMOs* (2012). Or maybe you have even seen the feature film *Percy* (2020), featuring Academy Award winner Christopher Walken and Golden Globe nominees Zach Braff and Christina Ricci on Amazon Prime. Either way, the story of Percy Schmeiser brings together all the elements of the conflict surrounding the right to patent "life" into one controversial story.

Let's go ahead and state the obvious: we were not on Percy Schmeiser's farm in 1996 and 1997, so there is really no way to know for certain what transpired. All we can go on is the case that was presented and ruled on in court. The premise of the case is pretty simple—Schmeiser claims that a storm during the canola harvest of 1996 blew canola pollen from a neighboring farm that was growing Monsanto varieties onto his own. Schmeiser saved his own seed and planted it the following year, "inadvertently" sowing his field with seeds that contained patented Monsanto genetic material that was modified with Roundup Ready technology.

Percy made a few pivotal claims:

1) He wasn't even aware that Monsanto's Roundup Ready canola seed existed.

2) He discovered that some canola along the field edge did not die when he sprayed the field ditches for weeds with Roundup.

3) The Monsanto seed contaminated his fields, resulting in devastating results for the farmer—lower yields the following season and ruining the seeds he had been developing for decades.

What about Monsanto's side of the story?

1) The company held informational meetings in Schmeiser's area and placed ads in local papers to coincide with the launch of their seeds, essentially making it hard for a farmer in the area to not know what Monsanto's Roundup Ready seeds were.

2) Other farmers in the area reported to Monsanto that Schmeiser was knowingly growing Monsanto seed on his farm without a license.

3) Independent testing that was approved by a court found over 90% of Schmeiser's crop contained Monsanto's Roundup Ready technology. A scientist testified in court that cross-pollination from contamination could not have occurred at the levels found on Schmeiser's farm.

The verdict? The case made it all the way to the Supreme Court of Canada, where the court ruled 5-4 in favor of Monsanto that Schmeiser intentionally used Monsanto's patented GMO seeds. However, they also ruled 9-0 that Schmeiser did not have to pay Monsanto for profits or damages because he did not receive any benefit from the technology.

What are your thoughts right now? Have you seen any of the movies mentioned earlier, because they certainly lean heavily to the side of Schmeiser being the victim. I think one of the most important parts of this story is the third point under Monsanto's claims. Even though Schmeiser claims he had been saving, cleaning, and planting his own seed for years, 98% of his field contained Monsanto's patented genetics. It is simply a statistical improbability that such high levels of "contamination" could have happened by accident and points very clearly to the explanation that Schmeiser intentionally saved the Roundup Ready canola seed. Why did he do it? It's hard to speculate on a person's motives, but the action is supported by the evidence in this case.

The opposing argument, without fail, pulls at the emotional strings that a case like Schmeiser vs Monsanto can strike. Who wouldn't want to cheer for the underdog against the giant global corporation? And how can a company claim rights to a farmer's livelihood, essentially owning his ability to plant seed that is physically in his hands at the end of harvest? Thinking back to the legislative measures that allow companies and individuals to patent plants and microbiology, and the ability to file for utility patents for the same materials, we discussed why this was important from an academic standpoint. But adding a personal element can often confound the idea of right and wrong.

Has any of this discussion swayed your opinion, or at least made you stop and think about your views on the ability to patent life? Let's add one more point to muddy the waters even further. It is relatively uncontroversial that genetic improvements from plant breeding have contributed significantly to yield gains in US crops. According to the USDA Economic Research Service, "50 percent or more of the overall yield gain for each crop [corn, soybean, cotton, and wheat] can be attributed to genetic improvements of plant varieties."[20] Yet, it takes roughly thirteen years at a cost of around $136 million (data from 2008-2012) [21] to bring a new plant biotechnology trait to market in the US. And that is just the trait. Considerable research, screening, and proliferation are performed at the major seed-producing companies to ensure that they are offering the best genetics possible along with the desired traits. If companies are not able to patent their intellectual and physical property so they can recoup their investment, what will incentivize them to continue performing this sort of research?

— So, Is It All Sunshine and Rainbows? —

We have discussed the rationale for broader acceptance of GMOs from a purely consumer safety viewpoint, but that's not to say there aren't areas for improvement. We should continually strive to improve our current systems and look for shortcomings that can be addressed, so let's turn our attention to some of the legitimate concerns about current practices in modern agriculture, particularly when it comes to proper stewardship of new technologies.

We closed the organic chapters with the claim that there is room for ideas from both organic and conventional agriculture to coexist and improve upon one another to advance our agricultural practices. The same is true for GMOs in agricultural production. Too often the agricultural industry views new technologies as the next silver bullet, putting undue pressure on the technology rather than focusing on systems-based solutions. This was certainly the case with the introduction and use of Roundup Ready crops.

Before glyphosate-tolerant crops were introduced in the mid-1990s, farmers needed to have a deep understanding of the crops they were planting, the crops that would be planted next, the types and species of weeds they were battling in a field, and the classes of chemistry available to them to get the job done. The same could be said for insect pests prior to the advent of Bt crops. But now we have GMOs, so we're all good, right? Not quite.

There is a lot we could unpack regarding pesticide use over time and the correlation with GMO use and adoption. The volume of herbicides applied on farms saw marked decreases in the late-1990s after glyphosate-tolerant crops were introduced to the market. However, if this trend was the feather in your hat to claim GMOs were saving the planet by reducing pesticide use, I'm sorry to tell you, but you probably need to find a new feather. Starting around 2005, herbicide usage began to pick back up after a short plateau in the early 2000s, and the increase continued at least through 2014.[22] While this trend is not necessarily positive, we should take note that it is strictly a measure of pounds of active ingredient applied per acre. Empirically, that is only one way to look at the data.

Another way to view the data, which more closely reflects the ideas presented in the first section of this book, focuses on the "quality" of the pesticides being used. Remember when we discussed how researchers have learned a lot about herbicides since the "Golden Age of Pesticides" after WWII? Since then, researchers have focused on increasing the efficacy of chemicals while simultaneously reducing their toxicity and persistence in the environment. Data summarized by the USDA ERS shows that the researchers and chemical companies have been successful in this quest as pesticide rates and chemical persistence were both cut in half between 1968 and 2008, and toxicity has been reduced to negligible levels over that same time.[23] Essentially, from a quality standpoint, the products we are using today are vastly superior to those we have used in the past. Now that is more like it!

But don't get too excited yet. I have another sobering reality for us. This one is courtesy of the International Herbicide-Resistant Weed Database. The oft-cited charts provided by this organization paint a pretty gloomy picture of what farmers are up against as the number of weed species that have evolved resistance to various herbicide modes of action continues to increase over time.[24] The older classes of chemistry, like ALS inhibitors and triazines, are definitely leading the way in the sheer number of resistant weed species, but a noticeable uptick in glyphosate-resistant weeds occurs right around 2000 and increases, nearly linearly, from that date on.

So, what should we make of this? Yes, the increase in glyphosate-resistant weeds starting around 2000 coincides with the introduction of glyphosate-tolerant crops in the late 1990s. But, it appears that weeds have evolved (in the case of PSII Inhibitors) or are still evolving (in the case of ALS Inhibitors) resistance to other herbicide modes of action at a faster rate than glyphosate.

This discussion is by no means meant to say we have stewarded the use of glyphosate and glyphosate-tolerant crops flawlessly . . . hardly. While Roundup

Ready was a great tool when it was first introduced, giving farmers the ability to apply a single herbicide over the top of glyphosate-tolerant crops and control every weed present in the field, it was misused and overused, and we are dealing with the repercussions now. Even when the early warning signs were at our door and glyphosate became less effective at killing weeds, we simply used higher rates or applied it more frequently. Our silver bullet, as it turns out, was not sterling.

It is often hard for us to think about longer-term benefits when we are not able to see the immediate rewards. Take the following example. Imagine that a new supplement promises to meet your daily nutritional needs and guarantees weight loss, regardless of your diet or exercise regimen. You decide to take it and the claims were true! Pizza for every meal, zero exercise, but you feel healthier than ever and the pounds are melting away. But wait. Medical experts are now saying that you need to add diet and exercise in addition to the supplement. You won't see any difference in results, and at least in the short term, you will feel the same and lose the same weight whether you diet and exercise or not; it will just require more effort and more money. But in the long-term, at some point, these experts believe that the supplement will eventually stop working. That sounds like it would be a hard sell, but that is essentially what farmers were presented with when it came to the benefits of Roundup and glyphosate-tolerant crops. So, before anyone starts giving farmers too hard of a time about the overuse and misuse of Roundup, would you have done it differently?

It is also important to note that crop traits and herbicides are simply tools. This is important because one of the arguments against GMOs and herbicide use is that they are creating "superweeds." These "superweeds" are "overrunning America's farm landscape," akin to a "plague [that] has spread across much of the country," which is "wreaking environmental havoc" and forcing farmers to "resort to more toxic weed-killers."[25] In case you think I am being unfair and cherry-picking the most egregious and emotionally charged language used to describe this topic, *all* of those quoted phrases in the last sentence were contained in just the opening paragraph of a document titled, "The Rise of Superweeds and What to Do About Them" by the Union of Concerned Scientists. I do not often refer to myself as a scientist, but while pursuing my PhD I was fortunate enough to be trained in scientific writing and have published peer-reviewed scientific journal articles. I can tell you with certainty that there is nothing "scientific" about the language I highlighted from this document (and I have some serious doubts about the use of the word "scientists" in the name of this organization, but I digress).

Not to continue to pick on this group or their publication, but the very next paragraph starts, "How did this happen? It turns out that big agribusiness, including the Monsanto Company, has spent much of the last two decades selling farmers products that would ultimately produce herbicide-resistant weeds." Before we get to the conflation of anti-corporate sentiment and anti-GMO arguments, herbicides that "produce herbicide-resistant weeds"? What?

This may seem like semantics, but it should be made clear that no one and no thing is *producing* herbicide-resistant weeds. To the casual observer, this language suggests that the herbicides are in some way *causing* a change in individual weeds, mutating the plants through some sort of toxic, sinister process. That's just not the way it works. There are two terms that are essential for this debate: mutations and selective pressure.

Mutations sound like a bad thing, and when it comes to genetic mutations, they often are. The term refers to a change in an organism's DNA that results in the alteration of a gene from its normal state. Even with more than three billion base pairs in the human genome, an alteration in just one of them can cause sickle-cell anemia if it occurs in a specific location.[26] However, changes in DNA can also alter genes that produce characteristics that are beneficial for an organism and allow them to be better adapted to a situation or environment. The final and most common fate of genetic change is, well, nothing. Most changes at the genetic level happen without any effect, but regardless of the outcome, genetic changes are *prolific* in nature.

Now let's discuss selection pressure through the lens of our new understanding of mutations, particularly as it relates to herbicides. Herbicides are applied to kill weeds by interfering with any number of specific and essential processes in the plant, referred to as the target site. However, naturally occurring genetic mutations can occur in the weeds that alter the target site, rendering the herbicide ineffective. If the weed containing the mutation survives to reproduction, the mutation that altered the target site could be passed on to the next generation. Continued use of the herbicide will continue to kill the "normal" weeds, but the naturally "mutated" weeds continue to survive, reproduce, and multiply until a large population of resistant weeds is established. This continued herbicide use that kills the normal weeds and allows only the mutated weeds to survive and reproduce is the selection pressure. While it could be stated that the herbicide is "causing" resistance in a population of weeds because of this selection pressure, it is not, in fact, mutating the weeds because of its application. Rather, it is simply "selecting" the plants that have naturally occurring mutations already in their genetic makeup. This, clearly, was not the intent of the manufacturers!

We highlighted positive examples of selection pressure at the very beginning of this section with the introduction of selective breeding. Another example of selection pressure I have seen referenced that provides some perspective to our current discussion revolves around the dandelion. We often see tall dandelions growing in fields, but in our yards, the dandelions are noticeably short, with the flowers and seed heads barely coming off the ground. This is because we have applied selection pressure to the dandelions with our lawnmowers so that only those that grow low enough to avoid being chopped up by the blades survive and reproduce. Using the same standard as the Union of Concerned Scientists, we homeowners are responsible for creating dandelion superweeds!

It is obviously a stretch to compare the significant ramifications of herbicide-resistant weeds with the innocuous shortening of dandelions in our yards, but both are examples of selection pressure. However, as I stated, herbicides and herbicide-tolerant crops are merely tools, just as mutations and selection pressures are inherent processes in nature. Weed management has been critical to increasing agricultural and food production, and we have unfortunately seen the result of our tools colliding with nature to create the significant challenges that we are facing today. But I reject the idea that we can look back with 20/20 vision and say that the use of these tools was ill-advised or done with malintent. Rather, I applaud the industry for continuing to learn how we can continue to address the struggle that comes with taming nature to produce the necessary output to sustain our population now and into the future.

— Is Anti-GMO Really Just Anti-Big Corporation? —

A few pages back, I made a passing comment about the conflation of anti-corporate sentiment and anti-GMO arguments. Is it possible that a lot of the passion and debate around GMOs is rooted in a resentment that permeates the public subconscious towards big corporations? It may even be exacerbated in this case because of an inherent sense of nostalgia that surrounds farming, possibly stemming from the naturalistic fallacy that has already been discussed. When it comes to anti-corporate sentiment in agriculture, Monsanto undoubtedly became public enemy number one, even earning the charming nickname "Monsatan."

As we previously mentioned, German pharmaceutical giant Bayer acquired Monsanto in 2018, taking ownership of the robust seed and technology portfolio of the company, but also the public's ire. We have attempted to provide a reasoned argument around the safety, and even necessity, of GMOs, but if we are to move forward and take advantage of new and promising technologies that are poised to help us address the monumental challenges facing global food production, we need to address the underlying issue of anti-corporate sentiment. We must determine if corporations are allies in our efforts or are they truly the boogeymen that the media and anti-GMO cohort make them out to be.

For the sake of simplicity, we are going to focus solely on Monsanto since they have drawn most of the public attention in this regard (and because they are no longer a legal entity!). There are many arguments and accusations levied against Monsanto, as well as attempts to provide a counter-narrative to the claims. While the undergirding theme is premised on the idea of global corporatism being inherently bad for society, the specific arguments have centered on the ruthless legal actions Monsanto has taken against farmers. Corollary arguments have been founded on the essential opposition to the ability to patent life, as well as the impact Monsanto's greedy policies have had on actual human life, specifically the epidemic of suicides in Indian farmers. Let's touch on each of these briefly so we can turn the page and begin to look at what the future holds for food biotechnology.

There is not a lot more that we can say about the "Monsanto aggressively sues farmers" claim other than what we have already covered in *The Right to Own . . . Life?* section, particularly regarding the Percy Schmeiser case. But what we unpacked about that case holds true for the entirety of the argument. The anti-GMO crowd claims that farmers are at risk of being liable for patent infringement from inadvertent contamination because Monsanto aggressively pursues legal action against farmers. The first part of the argument is patently false. In fact, Monsanto has never filed suit, and the company even went so far as to state that they never would file suit, for inadvertent contamination of their technology. While Schmeiser claimed that is what occurred in his case, it was ruled not to be the case in a court of law.

The second part of the argument, however, we would have to say is true. Monsanto did aggressively pursue legal action against farmers who *knowingly* infringed upon their patents by using their intellectual property without paying for the right to do so. In fairness, this tactic by Monsanto was actually good for the entire agricultural biotechnology industry even though it drew the negative attention from the public. But to those who have an issue with it, I would ask: in what other industry is it okay to infringe upon a person's or company's patent and expect to face no repercussions?

Perhaps one of the best comparisons is the pharmaceutical industry. Hundreds of millions of dollars are invested in research and development to bring new drugs to market. While we may want the drugs that are critical to the health and wellbeing of people (especially those that are lifesaving) to be free to those in need, it does not change the fact that the company took the initial risk to spend the money on R&D in hopes they would find a marketable drug. Eliminate the incentive structure to take these risks and the pharmaceutical pipeline dries up. There is always the risk that bad actors will take advantage of the system, but the free market is responsible for driving innovation. It is the same with Monsanto and every other manufacturer in agriculture. They spend hundreds of millions of dollars in R&D for new seeds, traits, chemicals, and technologies to drive progress in the industry. We cannot afford to lose that.

The final argument levied against Monsanto is particularly bold: that they are responsible for the high rates of suicide in the Indian farming population. One of the loudest voices promulgating this idea is Vandana Shiva, an Indian-born environmentalist who has been a leading opponent against Monsanto and GMOs. Her contention is that Monsanto introduced Bt cotton into India but failed to deliver on its promise to increase yields. Rather, farmers are left in financial ruin from high production costs and low yields, and the subsequent strain is driving farmers to commit suicide. It is a tragic story that was documented in the 2011 film *Bitter Seeds*. The only problem is that the story does not hold up to the facts. While nothing I am about to write is meant to diminish the tragedy of the loss of life experienced by the families in India, we must examine the data available surrounding the claims.

The first claim made by Shiva and others is that Bt cotton is somehow failing Indian farmers by not fulfilling the promise of increased yields. However, studies have shown cotton yields began to increase drastically after the turn of the millennium.[27] The documented yield increase actually starts before the adoption of Bt acres, but several commentators point to the widespread use of illegal Bt cotton varieties being planted before it was officially introduced in 2002.

Still, there is a level of nuance that we must consider if we are to have an honest conversation around this point. Charts that continue to track this data show yield leveling off while the adoption rate of Bt cotton continues to increase. There are also some discrepancies when yield per hectare is used instead of total country production. However, based on the data available to us, it is fair to say that Bt cotton has helped to increase cotton yield production in India.

But yield is only one part of the story as Bt cotton farmers have also realized higher net profits.[28] Bt cotton should not be credited solely for India's cotton yield improvements, or increases in income, because agricultural systems are complex. The Bt trait is only mitigating the impact of a single insect pest, meaning the benefit of the trait will be dependent upon the level of infestation of that pest. Fertility, water management, and other agronomic practices are a large part of driving yield in any crop, and it is no different for Indian cotton production. However, it is awfully hard to look at this data and come away with the conclusion that Bt cotton has somehow failed Indian farmers and caused increased suicides in the 1990s and 2000s.

What about the idea that Indian farmers are committing suicide at rates that vastly exceed that of the general Indian population or comparable populations elsewhere? This, after all, was the basis for Shiva's claim as she labeled Monsanto's Bt cotton "seeds of suicide." Between 1997 and 2005, it is estimated that one farmer committed suicide every thirty-two minutes,[29] which spurred the protest slogan of "Every 30 Minutes!" for the anti-GMO and anti-Monsanto activists like Shiva. The Center for Human Rights and Global Justice at New York University School of Law even published a report titled, "Every Thirty Minutes: Farmer Suicides, Human Rights, and the Agrarian Crisis in India."[30] This is a sobering number, and one that requires reflection on the part of all involved as to how to remedy the situation. But how can this be placed solely at the feet of Monsanto when the period for data collection begins in 1997, five years before Bt cotton was officially introduced into India, and at least three years before the seeds were first planted illegally?

Nevertheless, the report, like many others, make statements that insinuate Monsanto did not fulfill their promise to "allow the cotton crop to survive bollworm infestations" and result "in higher yields, decreased instances of crop failure, and, ultimately, in greater economic security for their families." The proof the authors provide for Monsanto's failure is highlighted in the next paragraph with an anecdotal example from an individual farmer that experienced a crop failure in 2008 "as a result of unpredictable weather conditions." I described earlier that the Bt trait protects the crop from a specific insect pest, one that causes significant yield losses in Indian cotton production. Why would it be the responsibility of Monsanto to reduce the impact of weather variability on a crop when their trait merely protects the crop against a single pest? This is like saying people that take Advil to reduce headaches are let down by Pfizer because they ended up being diagnosed with colon cancer. The two maladies are not related.

At the risk of sounding like a broken record, the statistic that one farmer is committing suicide in India every thirty minutes is an undeniable tragedy, but how does this relate to other metrics on the subject? The suicide rate in India for the general population was 10.5 per 100,000 people in 2002; the rate for farmers was estimated to be 15.8 per 100,000 peoplein 2001.[31] This translates to a roughly 50% increase in the farming community. Comparing that with the US, CDC data from 2016 show a suicide rate of 15.6 per 100,000 people for the general population.[32] Agriculture represents the third-highest industry in the US regarding suicide rate at 36.1 per 100,000 people, behind only Mining, Quarrying, and Oil and Gas Extraction (54.2 per 100,000) and Construction (45.3 per 100,000).[33] While this is not meant to be a competition and is certainly not one that is deserving of accolades, the suicide rate for the agricultural industry in the US is 131% higher than that of the general population, compared to the 50% increase in India.

An extensive review in the Journal of Epidemiology and Global Health evaluating the statistics and causations for farmer suicide in India concludes that "Bt cotton is unlikely to be an important factor" in the increased suicide rates in Indian farmers.[34] In fact, they found that while farmers using Bt cotton may have increased annual costs by approximately 15%, the reduced pesticides and increased yields result in a 58.2% higher net return per hectare. The predominant factor associated with farmer suicides was identified as indebtedness. They cite multiple studies that conclude the Indian farmers who commit suicides are overwhelmingly indebted, ranging from 86.5% to 98.5% of farmers in the studies. It may be easy to point the finger at a multinational corporation for a socioeconomic problem in a country, but if we are to remedy the situation, it would behoove us to identify the true cause.

The data points to the underlying financial situation for many rural farmers as the leading cause of farmer suicide in India. Rural Indian farmers, as is the case for many small-holder farmers in the developing world, have little access to institutional credit, as noted by the World Bank: "While India has a wide network of rural finance institutions, many of the rural poor remain excluded, due to inefficiencies in the formal finance institutions, the weak regulatory framework, high transaction costs, and risks associated with lending to agriculture.[35] Further, it is noted that the problem in India started with banking reforms that were initiated in the early 1990s. Banking became more competitive in India from foreign and private banks entering the market, and the result was fewer loans available to farmers because of the lower profitability and higher risk of agricultural loans.[36] Farmers have been forced to look for loans from private lenders that charge excessively high interest rates. The access to credit and high interest rate loans are more indicative of the suicide rates in farmers than the use of Bt cotton, or cotton production in general, but why let data and facts get in the way of a good narrative?

It is important to uncover the underlying motivations behind objections to technology so we can understand what is truly at stake. We are getting ready to talk about the coming biotech revolution that will help society meet the challenges we are all facing regarding food production. However, if we are held captive by a narrative that is based in fearmongering, like that of the attacks levied against corporations like Monsanto regarding farmer oppression and biotechnological warfare, we will be impotent to drive the change available to us. We must root ourselves in facts, even in the face of a captivating narrative when leading voices like Shiva claim that "Monsanto's Bt cotton has already pushed thousands of Indian farmers into debt, despair and death." Or Paul Ehrlich (remember him from earlier) declaring that Monsanto "killed those farmers in India," or alternative medicine doctor Joseph Mercola proclaiming that Monsanto "has blood on its hands," or even political figures like Prince Charles saying, "I blame GM crops for farmers' suicides."[37] Personally, I choose to believe in the mantra "facts over fear."

1 FoodChain (2021). *Non-GMO Project Verification: FAQ*. FoodChain ID. https://www.foodchainid.com/non-gmo/faq/

2 Lusk J (2018). *Want a non-GMO? How much more will it cost?* JaysonLusk.com. http://jaysonlusk.com/blog/2018/3/22/how-much-more-will-you-pay-for-non-gmo-food

3 USDA AMS (2021). *List of bioengineered foods*. United States Department of Agriculture Agricultural Marketing Service. https://www.ams.usda.gov/rules-regulations/be/bioengineered-foods-list

4 Bohl E (2019). *GMO-free marketing is deliberately misleading consumers*. Missouri Farm Bureau. https://mofb.org/gmo-free-marketing-is-deliberately-misleading-consumers/

5 Alliance for Science (2015). *Virus resistant papaya: Learn about how public-sector scientists saved papaya*. Cornell University. https://allianceforscience.cornell.edu/blog/2015/01/learn-about-how-public-sector-scientists-saved-papaya/

6 Alliance for Science (2014, August 18). *The story of rainbow papaya - Why public sector biotechnology research matters* [Video File]. YouTube. https://www.youtube.com/watch?v=CX7AqBOJS84

7 Mead FW (2017). *Asian citrus psyllid*. Featured Creatures. University of Florida IFAS. EENY-33. http://entnemdept.ufl.edu/creatures/citrus/acpsyllid.htm

8 Grafton-Cardwell E (2021). *Huanglongbing (HLB or citrus greening)*. Invasive Species. UC Riverside Center for Invasive Species. https://cisr.ucr.edu/invasive-species/huanglongbing-hlb-or-citrus-greening

9 Florida Department of Agricultural and Consumer Services (2019). *Florida citrus statistics 2017-2018*. USDA National Agricultural Statistics Service. https://www.nass.usda.gov/Statistics_by_State/Florida/Publications/Citrus/Citrus_Statistics/2017-18/fcs1718.pdf

10 Robinson A (2020). *HLB solution could be available in three years*. Agent West. https://agnetwest.com/hlb-solution-could-be-available-in-three-years/

11 NGP (2016). *GMO science*. Non-GMO Project. https://www.nongmoproject.org/gmo-facts/science/

12 AAAS (2012). *Statement by the AAAS Board of Directors on labeling of genetically modified foods*. American Association for the Advancement of Science. https://www.aaas.org/sites/default/files/AAAS_GM_statement.pdf

13 Ibid.

14 An Act to provide for plant patents, 71st Cong. §312 (1930). https://www.loc.gov/law/help/statutes-at-large/71st-congress/session-2/c71s2ch312.pdf

15 Pottage A and Sherman B (2007). Organisms and manufactures: On the history of plant inventions. *Melbourne University Law Review 22*. http://www.austlii.edu.au/au/journals/MelbULawRw/2007/22.html

16 Luther Burbank. Luther Burbank Home & Gardens. http://www.lutherburbank.org/about-us/luther-burbank

17 *Asgrow Seed Co. v. Winterboer*, 513 U.S. 179 (1995). https://www.law.cornell.edu/supct/html/92-2038.ZO.html

18 *Diamond v. Chakrabarty*, 447 U.S. 303 (1980). https://supreme.justia.com/cases/federal/us/447/303/

19 *J. E. M. Ag Supply, Inc. v. Pioneer Hi-Bred International, Inc.*, 534 U.S. 124 (2001). https://www.oyez.org/cases/2001/99-1996

20 Fernandez-Cornejo J (2004). The seed industry in U.S. agriculture: An exploration of data and information on crop seed markets, regulation, industry structure, and research and development. Agricultural Information Bulletin No. (AIB-786). USDA Economic Research Service. https://www.ers.usda.gov/publications/pub-details/?pubid=42531

21 Phillips McDougall (2011). *The cost and time involved in the discovery, development and authorization of a new plant biotechnology derived trait.* A Consultancy Study for Crop Life International. https://croplife.org/wp-content/uploads/2014/04/Getting-a-Biotech-Crop-to-Market-Phillips-McDougall-Study.pdf

22 Hellerstein D, Vilorio D, and Ribaudo M (Eds.) (2019). Agricultural resources and environmental indicators, 2019. *Economic Information Bulletin 208.* United States Department of Agriculture Economic Research Service. https://www.ers.usda.gov/webdocs/publications/93026/eib-208.pdf?v=8521.3

23 Fernandez-Cornejo J, Nehring R, Osteen C, Wechsler S, Martin A, and Vialou A (2014). Pesticide use in U.S. agriculture: 21 selected crops, 1960-2008. *Economic Information Bulletin 124.* United States Department of Agriculture Economic Research Service. https://www.ers.usda.gov/webdocs/publications/43854/46734_eib124.pdf

24 Heap I (2021). *The International Herbicide-Resistant Weed Database.* WeedScience.org. http://weedscience.org/Pages/Graphs/SOAGraph.aspx

25 UCS (2013). *The rise of superweeds – and what to do about it.* Policy Brief. Union of Concerned Scientists. https://www.ucsusa.org/sites/default/files/2019-09/rise-of-superweeds.pdf

26 Clancy S (2008). Genetic mutation. *Nature Education 1*(1), 187. https://www.nature.com/scitable/topicpage/genetic-mutation-441/

27 Kranthi KR and Stone GD (2020). Long-term impacts of Bt cotton in India. *Nature Plants 6*, 188-196. https://doi.org/10.1038/s41477-020-0615-5

28 Kathage J and Qaim M (2012). Economic impacts and impact dynamics of Bt (*Bacillus thuringiensis*) cotton in India. PNAS 109(29), 11652-56. https://doi.org/10.1073/pnas.1203647109

29 Dwivedi T (2020). *Farmer's suicides – An issue of great concern.* The Times of India. https://timesofindia.indiatimes.com/readersblog/hail-to-feminism/farmers-suicides-an-issue-of-great-concern-27472/

30 Center for Human Rights and Global Justice (2011). *Every thirty minutes: Farmer suicides, human rights, and the agrarian crisis in India.* NYU School of Law. https://chrgj.org/wp-content/uploads/2016/09/Farmer-Suicides.pdf

31 Berezow A (2017). *Vandana Shiva's myth busted: Monsanto didn't cause farmer suicides in India.* American Council on Science and Health. https://www.acsh.org/news/2017/01/07/vandana-shivas-myth-busted-monsanto-didnt-cause-farmer-suicides-india-10696

32 Stone DM, Simon TR, Fowler KA, Kegler SR, Yuan K, Holland KM, Ivey-Stephenson AZ, and Crosby AE (2018). Vital signs: Trends in state suicide rates – United States, 1996-2016 and circumstances contributing to suicide – 27 states, 2015. *Morbidity and Mortality Weekly Report 67*(22). Centers for Disease Control and Prevention. https://www.cdc.gov/mmwr/volumes/67/wr/pdfs/mm6722a1-H.pdf

33 Peterson C, Sussell A, Li J, Schumacher PK, Yeoman K, and Stone DM (2020). Suicide rates by industry and occupation – National violent death reporting system, 32 states, 2016. *Morbidity and Mortality Weekly Report 69*(3). Centers for Disease Control and Prevention. https://www.cdc.gov/mmwr/volumes/69/wr/mm6903a1.htm

34 Merriott D (2016). Factors associated with the farmer suicide crisis in India. *Journal of Epidemiology and Global Health 6*(4), 217-227. https://doi.org/10.1016/j.jegh.2016.03.003

35 Kloor K (2014). *The GMO-suicide myth.* Issues in Science and Technology. https://issues.org/keith-gmo-indian-farmers-suicide/

36 Sadanandan A (2014). Political economy of suicide: Financial reforms, credit crunches and farmer suicides in India. *Journal of Developing Areas 48*(4), 287-307. https://ssrn.com/abstract=2942490

37 Kloor K, The GMO-suicide myth.

Non-GMO Verified:
The Future is Bright

— The New Kid in Town —

It's finally time to turn our attention to the future of food production and what role genetic engineering will play. Our understanding of biology, genetics, and physiology has grown tremendously since the days of Mendel and his peas—from the first transfer of DNA from one bacterium to another to incorporating genes that encode for insecticides directly into crops. Science, it would appear, is up to the challenge of producing enough food to feed a growing population with increasing constraints on our natural resources and the need to do so in an environmentally friendly way. But what's next? Is there groundbreaking technology right around the corner? The answer is an emphatic YES!

Thus far we have only discussed one method for genetic engineering, and that is the use of *Agrobacterium* to carry a gene of interest into a plant that will be incorporated into the plant's DNA. While this is much more efficient than traditional crossbreeding methods, it is still not exact because scientists cannot direct exactly where the gene will be inserted. It is also the primary complaint of anti-GMO activists because of the perception that we are creating organisms that are "unnatural" in some way because they contain DNA that is not found in the organism in nature. What if there were a way around the inaccuracies of *Agrobacterium*-mediated genetic engineering and the perception of unnatural "transgenic" organisms? The leading technology poised to answer that question is CRISPR/Cas9.

So, what is this new technology with a weird name? CRISPR is an acronym that stands for clusters of regularly interspaced short palindromic repeats, which simply means that DNA has regions where there are areas of repeating nucleotides, or the A, G, T, and C letters of the DNA code, and spacers between these repeats. The Cas9 refers to a specific "*CRISPR-as*sociated" enzyme that acts like scissors directed by the CRISPR sequence. This process was identified in the early 2000s but first proposed as a method for gene editing by Jennifer Doudna and her colleague Emmanuelle Charpentier in 2012, an accomplishment that earned both women the Nobel Prize in Chemistry in 2020.[1]

Much like the use of *Agrobacterium tumefaciens* in plant breeding, the concept of CRISPR is taken from a naturally occurring phenomenon in nature. Scientists discovered that certain bacteria have "spacers" between repeating sections of DNA in the bacterial genome (the CRISPR region), and these were actually bits of viral DNA incorporated from viruses that had previously attacked the bacterium. These spacers served as memory banks for future attacks from the virus. Essentially, these short sequences of DNA are transcribed into what is known as "guide RNA strands" that match up with the viral DNA if it ever infects the bacteria again. The associated Cas9 protein (acting as the pair of scissors in the mechanism) then cuts the double-stranded DNA of the virus that the guide RNA recognized, destroying the viral DNA and preventing the virus from causing harm to the bacteria. This is some pretty advanced genetics, but it is really freaking cool!

Alright, so bacteria can recognize viral infections because of the DNA they have on record in the CRISPR region and cut it with Cas9 at a specific point, but how does this translate into a method for genetic engineering? The second part of the story involves the inherent ability of cells to repair DNA. If our DNA were unable to repair itself after being cut or damaged, this would spell bad news for us and our health, as it is estimated that an individual cell can suffer from nearly *1 million* DNA changes every day![2] To combat this, our cells possess the ability to "glue" DNA that has been cut back together, either by (1) sticking the two cut ends back together, known as non-homologous end joining (NHEJ), albeit with minor changes to the sequence, or (2) integrating new DNA at the site of the breakage, known as homology-directed repair (HDR). For the purposes of genetic engineering, both pathways can be important. Repair by NHEJ often results in a mismatch of the original DNA sequence, which could cause a gene to be "knocked out" if it occurs at a specific location. Alternatively, the addition of novel DNA through the HDR pathway offers a mechanism to introduce new genes into the study organism.

Researchers are investigating this ability to cut DNA at specific locations in a target genome to drive genome editing. Full transparency, there is still a lot of work to be done. While the CRISPR guide RNA will target an exact sequence of DNA, this sequence can occur at multiple locations within the genome of the organism being researched. Take humans, for example, who have over 3.2 billion base pairs in our genome. The chance that a twenty-nucleotide sequence, the target sequence of the guide RNA, occurs in more than one location in our entire genetic blueprint is relatively high (0.3% chance according to my rudimentary calculations assuming for completely random assortments of base pairs across the entire genome). Researchers must use our understanding from genome mapping to ensure that they are able to target CRISPR cuts and edits exactly—and only where they wish to make an edit. While the consequences are less dire in plant breeding (as edited plants can be grown to determine if any unintended genetic alterations have resulted in negative manifestations), the use of gene-editing in humans has a much smaller margin of error.

The second issue revolves around the repair process. The NHEJ is useful when researchers are looking to "knock out" a gene of interest, but what about adding a new gene? While researchers can supply cells with the desired DNA fragment to be used in the repair process, the use of the desired sequence does not occur 100% of the time. As a result, scientists must continue to work on improving the use of the selected DNA fragment in the repair process.

So, while we have identified two areas that need to be refined as we learn more about the use of CRISPR in gene editing, the technology is undoubtedly opening a world of possibilities for researchers across multiple fields. One of the primary reasons is that the use of CRISPR can be implemented in living organisms, not just on individual cells. Some of the earliest work has focused on therapies with blood cells and stem cells, like those found in bone marrow. Examples include those of two clinical patients, one afflicted with sickle cell anemia and the other with the blood disorder β-thalassemia, both of which affect the protein hemoglobin that carries oxygen around the body. Each patient has had complete transfusions of CRISPR-edited blood cells and have demonstrated positive early results.[3] In another first that involves directly injecting CRISPR gene-editing machinery into a patient, researchers are hoping to help a patient suffering from Leber congenital amaurosis, a genetic disorder that progressively destroys the light-sensing cells in the retina, to regain and improve their vision by restoring a critical protein in the eye.[4]

CRISPR gene-editing is also an active area of research in plant breeding. One of the benefits that we haven't mentioned yet is a phenomenon known as multiplexing. While many human diseases result from a single gene mutation, or even nucleotide change within a gene, many characteristics we breed for in plants result from multiple genes. Using CRISPR, scientists can modify multiple targets simultaneously to produce the desired results within a single generation instead of multiple generations with selective breeding. Yield and stress tolerance characteristics are examples of desirable qualities that are influenced by multiple genes, referred to as quantitative trait loci (QTL). Researchers are currently conducting studies in rice where modifications improved yield-related traits like dense and erect panicles and reduced plant height; in soybeans, where the photoperiod was delayed resulting in increased vegetative growth; and in tomatoes, where multiple genes were modified to cause increased early flowering that resulted in early flower production and yield.[5] Additionally, the technology has been used to improve the nutritional content in crops, introduce resistance to pathogens and herbicides, and increase tolerances to abiotic stresses like cold, drought, and salt.[6]

Our understanding of CRISPR is still in its infancy, but the relatively low cost and ease by which the process can be employed are encouraging for researchers around the world. Plant breeders can now work to incorporate multiple genes into single generations, vastly shortening the time to create new varieties. Additionally, because researchers can more accurately target where genetic modifications will occur, breeding programs can produce new plant varieties with the desired characteristics without unwanted traits resulting from errant genetic changes.

— The Regulatory Challenges —

I will admit that we didn't do the topic of CRISPR-mediated gene editing justice with our cursory overview. I would encourage those of you who find genetics interesting to look further into the technology and advancements being made. It is truly fascinating and encouraging at the same time. But most importantly for our conversation is the path towards acceptance and adoption of this technology when it is ready for commercialization.

To realize the benefits this technology can offer, it needs to be not only accepted, but welcomed into the public sphere. Regulations are already playing a big part and it is getting messy with how governmental regulatory bodies and countries are approaching the issue. Are plants modified through gene-editing techniques (like CRISPR) genetically modified organisms if they contain no DNA from other organisms? What, if any, additional testing is required to evaluate allergenicity, toxicity, or off-target effects? Can these plants be introduced to the market at all? These are the types of questions currently being debated, and the answers will ultimately decide the future of this technology and the good it can do.

We spent a significant amount of time in the organic section of this book on the regulatory processes that are involved with the agrochemical industry. This was to provide some level of comfort that the pesticide industry has ample oversight to ensure that new products and technologies are safe for public consumption. While we won't go into that level of detail regarding GMOs, it should be noted that this industry is carefully regulated as well. Before the FLAVR SAVR tomato hit store shelves, the US government had already established the "Coordinated Framework for the Regulation of Biotechnologies" in 1986. Amended in 1992 and most recently updated in 2017, the framework ensures that regulations continue to evolve with the ever-advancing field of genetics.

The EPA has oversight on the pesticides used on GMO crops and the pesticides produced by GMO crops, called plant-incorporated pesticides (PIPs). The FDA ensures that the same safety standards that are applied to all other foods are also met by GMOs. And lastly, the USDA evaluates the potential for GMOs to have off-target effects that could harm other plants. There is no shortage of oversight on genetically engineered crops! Let's turn back the clock a little bit to get some history on how and why GMOs are regulated the way they are.

In the late 1980s, the value of biological diversity and the need to protect it on a global scale was gaining international recognition. The United Nations Environment Programme (UNEP) led the charge, eventually forming the Convention of Biological Diversity (CBD) at the 1992 UN Conference on Environment and Development, more commonly referred to as the Rio Earth Summit. According to the CBD's website, the convention "was inspired by the world community's growing commitment to sustainable development," and focused on three objectives: 1) "the conservation of biological diversity," 2) "the sustainable use of its components," and 3) "the fair and equitable sharing of benefits arising from the use of genetic resources.[7] You may be wondering what this has to do with our current discussion . . . I'm glad you asked.

In 2000, the CBD adopted the Cartagena Protocol on Biosafety, or Biosafety Protocol for short, which is an international agreement meant "to ensure the safe handling, transport and use of living modified organisms (LMOs) resulting from modern biotechnology that may have adverse effects on biological diversity, taking into account risks to human health."[8] We mentioned this briefly when discussing Golden Rice, but the important takeaway is that this global pact established a framework by which all participating nations would agree to protocols and guidelines regarding LMOs, or as we have been referring to them, GMOs. "Key Protocol Issues" were identified and included topics such as handling, transport, packaging, identification, information sharing, monitoring, reporting, risk assessment, and risk management, to name a few.[9]

Ten years after the Biosafety Protocol, another protocol was adopted at the 2010 conference held in Nagoya, Japan, titled the Nagoya Protocol on Access to Genetic Resources and the Fair and Equitable Sharing of Benefits Arising from their Utilization (ABS) to the Convention on Biological Diversity. Wow, quite a mouthful! The purpose of the Nagoya Protocol was to implement the third objective of the CBD: the fair and equitable sharing of benefits arising out of the utilization of genetic resources. The CBD claims the Nagoya Protocol was important because it established "more predictable conditions for access to genetic resources" and helped "to ensure benefit-sharing when genetic resources leave the country providing the genetic resources."[10]

The language used in the original convention and the subsequent protocols conveys a very cooperative attitude among the partner countries, all working together for the benefit of the common good. As of 2016, every nation on the planet had ratified its membership to the CBD except two: the Vatican (or Holy See) and the United States—196 countries in all. Even though the number of nations that have ratified the Biosafety Protocol (176) and Nagoya Protocol (129) is less, most nation-states have committed to these agreements. So why has the United States held out on joining this global coalition? The answer is not clear-cut, but perhaps unsurprisingly, it has been infused with politics for nearly thirty years. One of the most pointed justifications for not joining the Convention came in 1992 from then-Senate Minority Leader Bob Dole. Providing comments during a Senate debate, Dole articulated his case in a lengthy statement. Portions of that statement are provided below:

"Environmental laws and regulations governing nearly every aspect of life in America are stronger in the United Stated than they are in any other country in the world. We have laws on air emissions, water discharges, filling and dredging wetlands and waterways, disposal of every type of waste from common household garbage to toxic chemicals to radioactive waste... The United States wants to have a cooperative agreement whereby all nations of the world commit themselves to undertake the same type of aggressive environmental controls that the United States has taken. Conversely, the Third World has viewed these negotiations as a cash cow. For a price, they have said, we might be able to interest them in being concerned about the environment... Ask the American taxpayer the real question: Do you favor spending hundreds of millions, if not billions, of your tax dollars to foreign countries to try to interest them in the environment? Or, do you favor taking a tough stand, demand that all nations follow the lead of the United States in cleaning the air, the water, protecting forests and species, and eliminating chemicals...?"[11]

While Bill Clinton did sign on to the Convention in 1993, the vote to ratify the United States' membership was struck down in Congress the following year.

In the minds of many policymakers and voters in the US, there is often broad agreement with Dole's statements that we, as a country, are leading the way in both scientific discovery *and* environmental protection. As proof, we can point to the numerous policies that were highlighted during our discussion on pesticides earlier in the book. The Federal Insecticide, Fungicide, and Rodenticide Act (FIFRA), the Endangered Species Act (ESA), the Clean Air and Clean Water Acts (CAA and CWA)—all pieces of legislation that were implemented to protect both our citizens and the environment. We even discussed the numerous agencies that are involved in regulating GMOs and the different roles the EPA, USDA, and FDA all play to ensure the safety of modern biotechnological advancements in food production.

Perhaps the most important point of all is the fact that the United States took these steps independent from an overarching obligation to a non-binding global agreement. For better or worse, an underlying belief in American exceptionalism and hesitancy towards globalism drives much of America's international policy, particularly when it comes to large, international agreements that are perceived as the United States getting "the short end of the stick." This independence is also what drives the commitment of the United States to support the development and advancement of GMOs and other agricultural advancements.

On the other side of that coin lies the European Union (EU), whose member states are all ratified parties to the CBD, the Biosafety Protocol, and the Nagoya Protocol. Adhering to the precautionary principle regarding GMOs, the EU has adopted a risk-averse stance towards modern biotechnology and has imposed strict regulations on GMOs. A policy brief by the Purdue Policy Research Institute describes the EU regulatory scheme, highlighting an extensive authorization process, the labeling of GMOs and assignment of a unique identifier, and a multi-step process to approve cultivation of GM crops, starting with the member states and ending with the final approval by the EU.[12] Even with these restrictive policies, member states can opt for even stricter regulations within their boundaries. According to the brief, governments that adopt more stringent policies, like the EU, cite public fear or concern as the primary reason for their decision.

This means countries are making policy decisions on subjective public sentiment rather than on objective scientific evidence, like that presented in the book *Genetically Engineered Crops: Experiences and Prospects* under the chapter titled "Human health effects of genetically engineered crops."[13] The comprehensive review presents findings from studies that investigated health concerns related to GM food consumption, ranging from allergenicity to gastrointestinal tract diseases. They even reviewed cancer incidence, comparing the trends in incidence of multiple cancer types in men from the United States and the United Kingdom, essentially drawing the comparison between a country that has widely adopted GM foods to one where GM foods are generally not consumed. The conclusion? "There is no obvious difference in the patterns that could be attributed to the increase in consumption of GE foods in the United States."

The question now at hand is how are nations going to approach the regulation and acceptance of newer genome editing techniques like CRISPR that we have already reviewed? If we view the United States on one end of the spectrum (with greater advocacy for GMOs to date) and the European Union on the other end (with strict regulations and outright bans of GMOs in some instances), most other countries fall somewhere in the middle. This mentality appears to be carrying over to the regulation of genome editing as well.

To be clear, genome editing through techniques like CRISPR do not automatically result in what we typically envision as a GMO. You may remember that earlier in this section we provided the definitions of various terms used in this field. Genetic modification is merely a selective pressure that causes desired outcomes in the offspring of an organism. We gave examples as innocuous as the labradoodle to highlight the fact that genetic modification is part of any breeding program, past, present, or future. The term genetically modified organism, however, refers specifically to an organism whose genetic material has been altered by genetic engineering. Even this definition does not truly define the practice that most people have an issue with regarding this subject. According to the Non-GMO Project, it is the use of "genetic engineering or transgenic technology" that results in a GMO, and specifically that it creates combinations of "genes that do not occur in nature or through traditional crossbreeding methods."[14] It is not until we start talking about transgenics, or the insertion of DNA from one organism into another, that people really begin to take a step back.

But CRISPR and other genome editing techniques may or may not result in transgenic organisms. If no additional DNA fragment is supplied, the DNA breakage caused by the CRISPR/Cas9 complex will be repaired by the normal genetic machinery of the cell. The result could be a deletion, insertion, or nucleotide substitution depending on how the repair process is undertaken. So, if DNA breakages occur naturally, and the repair is innate to the organism, then the only way this process differs from what happens in nature is the specificity of the site where the breakage occurs. It would not meet the criteria of even the Non-GMO Project! Why would it need to be regulated or scrutinized any differently than the unadulterated version of the organism in question?

This was the reasoning behind the USDA's decision not to regulate genetically edited organisms if no transgene was inserted. The agency declared that in these circumstances, genome editing is no different than conventional breeding. Australia agreed and amended their Gene Technology Act in 2019 to exclude organisms from being regulated as GMOs that were modified by CRISPR/Cas9 and similar methods. Other countries like Brazil, Argentina, and Chile have followed this rationale and evaluated genetically edited organisms on a case-by-case basis, exempting products that do not contain a transgene from regulation. It is worth noting that none of these countries are party to the Biosafety Protocol that defined living modified organisms and are therefore not "required" to be in full compatibility with the regulations of the protocol.[15]

As you can imagine, the European Union has not followed this logic. In 2018, the Court of Justice of the European Union (ECJ) ruled that genetically edited organisms will be subject to the same regulations and restrictions as GMOs per the 2001 EU GMO directive. While the European Union is relatively unique in their overt opposition to genetically edited organisms, other countries like New Zealand, Norway, South Africa, and Sweden have taken a similar approach to regulate them as GMOs, keeping them in compliance with their obligations as parties to the Biosafety Protocol.

It is interesting to note that the 2001 EU GMO directive does allow the use of mutagenesis to genetically modify organisms. We mentioned this method previously, but just to reiterate, mutagenesis is the process by which an organism is exposed to high doses of radiation to cause mutations in its DNA—the goal being that some radiation-induced mutations will occur in genes that result in a favorable outcome. It would seem logical that a main argument by scientists and plant breeders opposed to the ECJ ruling is the fact that gene-editing techniques like CRISPR should be considered mutagenesis because only small changes to DNA are occurring, like those that occur with the approved irradiation practices. The only difference is that it is more precise and, it may go without saying, but . . . **you don't have to use harmful levels of mutation-inducing radiation!** Unfortunately, logic does not appear to be a factor in this dispute and plant breeders in EU nations will have to sit on the sidelines as other developed nations continue to advance in the field of biotechnology to address the challenges facing our planet. But even more troubling, nations that trade with the EU will be hampered by the restrictive regulations of the region, which is already impairing the advancements in the field.

food for thought...

A final point to ponder. We have discussed the rationale for considering newer genome editing techniques independently from traditional genetic engineering techniques. However, the discussion to this point has focused on genetically edited organisms that do not contain a transgene. This concept is rooted in our understanding of traditional genetic engineering techniques and should be reevaluated as well. The concept of a transgene has historically been the literal isolation of a piece of DNA from an organism's genome, replication of that DNA, and then the insertion of that DNA into a new organism. However, new methods that enable the synthesis of strands of nucleic acids from scratch are changing this process, and quite possibly the entire idea of the transgene.

If the contention with transgenes is the source of the genetic material, then the new technologies should alleviate that concern. DNA is simply an ordered string of nucleotides, which themselves are simply small chemical compounds. With advancements in biotechnology, the nucleotides no longer need to be isolated from actual DNA; they can be synthesized the same as other chemicals, like pharmaceutical drugs. The comparison to pharmaceuticals may be a relevant one, because one of the promises of CRISPR and gene editing techniques is gene therapy. Just as we take medicines (AKA chemical compounds) to cure our maladies, CRISPR can deliver strands of nucleotides (AKA chemical compounds) to express therapeutic genes that can help cure more serious diseases. So, if we are just putting a synthesized chemical compound into our body to cure an illness, are we really talking about transgenes anymore?

1 Fernholm A (2020). Genetic scissors: *A tool for rewriting the code of life*. The Nobel Prize in Chemistry 2000. NobelPrize.org. https://www.nobelprize.org/prizes/chemistry/2020/popular-information/

2 Clancy S (2008). DNA damage & repair: mechanisms for maintaining DNA integrity. *Nature Education 1*(1):103. https://www.nature.com/scitable/topicpage/dna-damage-repair-mechanisms-for-maintaining-dna-344/

3 Le Page M (2020). *Three people with inherited diseases successfully treated with CRISPR*. NewScientist. https://www.newscientist.com/article/2246020-three-people-with-inherited-diseases-successfully-treated-with-crispr/

4 Stein R (2020). *In a 1st, scientists use revolutionary gene-editing tool to edit inside a patient*. Health Shots from NPR. https://www.npr.org/sections/health-shots/2020/03/04/811461486/in-a-1st-scientists-use-revolutionary-gene-editing-tool-to-edit-inside-a-patient

5 ISAAA (2021). *Pocket K No. 54: Plant Breeding Innovation: CRISPR-Cas9*. International Service for the Acquisition of Agri-biotech Applications. https://www.isaaa.org/resources/publications/pocketk/54/default.asp

6 El-Mounadi K, Morales-Floriano ML, and Garcia-Ruiz H (2020). Principles, applications, and biosafety of plant genome editing using CRISPR-Cas9. *Frontiers in Plant Science 11*(56). https://doi.org/10.3389/fpls.2020.00056

7 CBD (2021). *History of the convention*. Convention on Biological Diversity. https://www.cbd.int/history/

8 CBD (2021). *The Cartagena Protocol on Biosafety*. Convention on Biological Diversity. http://bch.cbd.int/protocol/

9 CBD (2021). *Text of the Cartagena Protocol on Biosafety*. Convention on Biological Diversity. http://bch.cbd.int/protocol/text

10 CBD (2021). *About the Nagoya Protocol*. Convention on Biological Diversity. https://www.cbd.int/abs/about/default.shtml/

11 Blomquist RF (2002). Ratification resisted: Understanding America's response to the Convention on Biological Diversity, 1989-2002. *Golden Gate University Law Review 32*(4). http://digitalcommons.law.ggu.edu/ggulrev/vol32/iss4/5

12 Paine J (2018). Global GMO policy. *Purdue Policy Research Institute (PPRI) Policy Briefs 4*(1). https://docs.lib.purdue.edu/gpripb/vol4/iss1/4

13 National Academies of Sciences, Engineering, and Medicine (2016). Human health effects of genetically engineered crops. In *Genetically Engineered Crop: Experiences and Prospects*. Washington, DC: The National Academies Press. https://doi.org/10.17226/23395.

14 NGP (2016). *What is a GMO?* Non-GMO Project. https://www.nongmoproject.org/gmo-facts/what-is-gmo/

15 El-Mounadi K, Morales-Floriano ML, and Garcia-Ruiz H, Principles, applications, and biosafety of plant genome editing using CRISPR-Cas9.

Vision of Hope

Anyone in the agriculture industry can parrot back the oft-quoted challenge facing agriculture—we must feed 9 to 10 billion people by 2050 on the same amount of land with limited resources and a focus towards being better stewards of the environment. (For reference, as of October 2021, the current population was just shy of 7.9 billion.) Seems like a tall order, but humans have been meeting the challenges that have faced society for millennia, and I see no reason why this challenge will be any different. We have already made tremendous advancements in the area of agriculture and food production, we have increased our awareness of the impact our actions can have on our shared resources, and we continue to innovate in the field of biotechnology. Yes, it may still seem daunting, and yes, it may even seem like we are in uncharted territory because of the complexity and global scale of the task at hand. But what challenge doesn't seem unprecedented when you are in the midst of it?

It is worth noting that there are aspects of this challenge that we haven't even covered in this text, like food waste, allocation, and policy. However, for the purpose of this book, we have focused on the different systems and methodologies involved in agricultural production. Even if we work towards improving food distribution and waste, we will still need to sustainably increase food production across the planet, especially in areas suffering from food insecurity.

I believe the only barrier that could impede our success in meeting this challenge would be the reluctance of individuals to find solutions rooted in common ground. But as we just mentioned, this barrier is not without precedent. Scientific innovations have been the engine for societal progress for millennia, yet with most innovations, especially those that could be considered a paradigm shift in understanding, they seem to be accompanied by an obligatory opposition to the finding.

Pythagoras' claim in the 6th Century BC that the Earth was round, supported by Aristotle nearly two hundred years later, was not truly accepted until explorers like Christopher Columbus set sail across the open seas. Nicolaus Copernicus' concept of heliotropism that posited the Earth is not, in fact, stationary but rather revolves around the Sun was met with vehement opposition. His compendium *On the Revolutions of the Heavenly Spheres*, published in 1543, was deemed heretical by the Catholic church, and Galileo was charged with heresy for supporting the doctrine.

Even the positive effects of handwashing were rejected at first. In the mid-1800s, Hungarian obstetrician Ignaz Semmelweis observed that mortality rates of mothers from puerperal fever following childbirth were greater when medical students assisted doctors than when midwives did. He realized that many of the medical students came directly from performing autopsies in the morgue and that they were likely transmitting something from the cadavers that was causing the disease. He implemented the practice of handwashing with chlorinated lime solutions and saw drastic reductions in mortality rates. However, this was several years before Louis Pasteur would confirm the germ theory, and Semmelweis' findings were deemed to be insufficiently supported by any known scientific explanations. Although he is now known as the "savior of mothers," during his life he was ridiculed for his theory and committed to an asylum.

Each of these examples illustrate the long-established dismissal, and even condemnation, of scientific discovery that has occurred throughout history because of pervading worldviews—sometimes from the public, sometimes from places of authority, and sometimes from the scientific community itself. While acceptance is the ultimate outcome in each case, the time required varies greatly from years to centuries before these visionaries were ultimately vindicated.

While it may seem maddening to look back in history with our current understanding of the science, healthy skepticism in its truest form will always result in greater understanding and advancement. Discovery without dispute would be nearly as detrimental as absolute resistance to invention. Taking the previous example of Copernicus, had it not been for his skepticism of the geocentric model established by Ptolemy nearly 1,400 years earlier, we may still believe the Earth is the center of the universe. But even his theory, further supported by Galileo, was flawed as it assumed the planets orbited the Sun in a circular motion. It took further observational data for Johannes Keppler to discover that planets actually move in elliptical orbits around the Sun.

So, where does that leave us with the current challenge at hand—how to feed a growing, global population? Specifically, how will we find common ground around the use of pesticides and GMOs in our agricultural systems? Or put another way, how will the growing demand and advocacy for organic and non-GMO food impact our solutions? Throughout this text, I have attempted to lay out the case for conventional agricultural systems because I believe they are the best option we have to meet the challenge.

Perhaps a more accurate way to phrase this is that I do not believe organic agricultural systems, in their purest sense, are the solution. This is an important distinction because it leaves room for the principles of organic agriculture to play a part in addressing the challenge. Actually, I would argue that they already have, from raising awareness around the hazards of pesticide use when persistence and selectivity were not critical selection criteria, to the introduction of alternative management practices like cover crops and intentional crop rotations. However, for common ground to be had, we must agree that pesticide use in itself is not inherently bad. Let's continue to strive for practices that promote more sustainable, systems-based approaches to farming. But let's also agree that the use of chemicals, which have been evaluated under incredibly stringent regulatory processes for safety, are still essential to minimize losses from pests. Or that the precise application of fertilizers, which have been processed and manufactured to ensure we can meet the nutritional needs of our crops, is critical to maximizing productivity. When the body of science concludes that these products are safe, let's not let ideology get in the way of providing food to those who need it most.

The debate over GMOs is more contentious, but I believe that is likely because the science regarding this discipline is still in its relative infancy. After all, it was only 1953 when Watson and Crick announced their discovery of the double-helix structure of DNA. Are we now in that obligatory *skeptical phase* that seems to accompany many of the scientific paradigm shifts like those we have just described? Possibly, but I wonder, is the skepticism more like the Catholic church's rejection of heliocentrism because it rejected a deeply held belief, or that of Johannes Kepler who had observational data that refuted a current theory?

I don't mean to sound hyperbolic, but it seems to me that the current opposition we are seeing to GMOs echoes that of the Catholic church declaring heliocentrism to be heresy. That is not in reference to the average consumer who makes a choice in the grocery aisle based on what they hear on social media. No, that is a reference to the activists who promulgate the anti-GMO narrative, relying on emotional appeals that the technology is in some way "unnatural."

The body of data regarding GMOs, and the overwhelming consensus from the scientific community, has determined that the technology is safe. And, as was the case with the use of pesticides, I believe we must come to an agreement that we should utilize all of the tools at our disposal to meet the challenge in front of us. Our current rate of crop yield improvements will not likely get us there, so it is going to require innovations in the field of biotechnology to change the rate of gain. It would be a travesty if we allow our feelings about technology (which are only possible because of the food security with which we are privileged) stifle innovation. We have already seen an example of this with the successful campaign against Golden Rice. It is incumbent upon all of us to investigate, think, and rationalize our thoughts. As Ayn Rand writes in *Atlas Shrugged*, "Devotion to the truth is the hallmark of morality; there is no greater, nobler, more heroic form of devotion than the act of a man who assumes the responsibility of thinking."

As we part ways, I want to thank you for taking the time to learn more about agriculture from a perspective you may not hear every day. We are living in a unique time where we have unlimited access to information, simply by pulling out our phones or uttering the word "Alexa." But this also poses a unique challenge. While I think an argument could be made that all information is biased in some way, we are dealing with the phenomenon today that the *delivery* of the information is increasingly biased. Search engines and social media platforms use algorithms to prioritize search results and news feeds based on trending topics, determined by what you like to see and what keeps you on the site longer. How much of what we know today is based on reality, or is it simply our own personal reality created by the information that has been curated for us?

As American historian Daniel J Boorstin stated:

"The greatest obstacle to discovery is not ignorance—it is the illusion of knowledge."